CO_2 Abatement and Economic Structural Change in the European Internal Market

AF757625

Contributions to Economics

Peter R. Haiss
Cultural Influences on Strategic Planning
1990. ISBN 3-7908-0481-9

Manfred Kremer/Marion Weber (Eds.)
Transforming Economic Systems: The Case of Poland
1992. ISBN 3-7908-0585-8

Marcel F. van Marion
Liberal Trade and Japan
1993. ISBN 3-7908-0699-4

Hans Schneeweiß/
Klaus F. Zimmermann (Eds.)
Studies in Applied Econometrics
1993. ISBN 3-7908-0716-8

Gerhard Gehrig/Wladyslaw Welfe (Eds.)
Economies in Transition
1993. ISBN 3-7908-0721-4

Christoph M. Schneider
Research and Development Management: From the Soviet Union to Russia
1994. ISBN 3-7908-0757-5

Bernhard Böhm/Lionello F. Punzo (Eds.)
Economic Performance
1994. ISBN 3-7908-0811-3

Michael Reiter
The Dynamics of Business Cycles
1995. ISBN 3-7908-0823-7

Michael Carlberg
Sustainability and Optimality of Public Debt
1995. ISBN 3-7908-0834-2

Lars Olof Persson/Ulf Wiberg
Microregional Fragmentation
1995. ISBN 3-7908-0855-5

Andreas Foerster
Institutional Analysis of Development Administration
1995. ISBN 3-7908-0853-9

Ernesto Felli/Furio C. Rosati/
Giovanni Tria (Eds.)
The Service Sector: Productivity and Growth
1995. ISBN 3-7908-0875-X

Giuseppe Munda
Multicriteria Evaluation in Fuzzy Environment
1995. ISBN 3-7908-0892-X

Boris Maurer
R & D, Innovation and Industrial Structure
1996. ISBN 3-7908-0900-4

Giovanni Galizzi/Luciano Venturini (Eds.)
Economics of Innovation: The Case of Food Industry
1996. ISBN 3-7908-0911-X

David T. Johnson
Poverty, Inequality and Social Welfare in Australia
1996. ISBN 3-7908-0942-X

Rongxing Guo
Border-Regional Economics
1996. ISBN 3-7908-0943-8

Oliver Fratzscher
The Political Economy of Trade Integration
1996. ISBN 3-7908-0945-4

Ulrich Landwehr
Industrial Mobility and Public Policy
1996. ISBN 3-7908-0949-7

Arnold Picot/Ekkehard Schlicht (Eds.)
Firms, Markets, and Contracts
1996. ISBN 3-7908-0947-0

Karin Peschel (Ed.)
Regional Growth and Regional Policy Within the Framework of European Integration
1997. ISBN 3-7908-0957-8

Thorsten Wichmann
Agricultural Technical Progress and the Development of a Dual Economy
1997. ISBN 3-7908-0960-8

Ulrich Woitek
Business Cycles
1997. ISBN 3-7908-0997-7

Michael Carlberg
International Economic Growth
1997. ISBN 3-7908-0995-0

Massimo Filippini
Elements of the Swiss Market for Electricity
1997. ISBN 3-7908-0996-9

Giuseppe Gaburro
Ethics and Economics
1997. ISBN 3-7908-0986-1

Frank Hoster · Heinz Welsch
Christoph Böhringer

CO_2 Abatement and Economic Structural Change in the European Internal Market

With 5 Figures
and 83 Tables

Physica-Verlag
A Springer-Verlag Company

Series Editors
Werner A. Müller
Peter Schuster

Authors
Dr. Frank Hoster
Institute of Energy Economics
University of Cologne
Albertus-Magnus-Platz
D-50923 Köln, Germany

Prof. Dr. Heinz Welsch
Department of Economics
University of Oldenburg
Ammerländer Heerstr. 114–118
D-26129 Oldenburg, Germany

Dr. Christoph Böhringer
Institute of Energy Economics
and the Rational Use of Energy
University of Stuttgart
Heßbrühlstr. 49a
D-70550 Stuttgart, Germany

ISBN 3-7908-1020-7 Physica-Verlag Heidelberg

Cataloging-in-Publication Data applied for
Die Deutsche Bibliothek – CIP-Einheitsaufnahme
Hoster, Frank: CO_2 abatement and economic structural change in the european internal market; with 83 tables / Frank Hoster; Heinz Welsch; Christoph Böhringer. – Heidelberg: Physica-Verl., 1997
(Contributions to economics)
ISBN 3-7908-1020-7

This work is subject to copyright. All rights are reserved, whether the whole or part of the material is concerned, specifically the rights of translation, reprinting, reuse of illustrations, recitation, broadcasting, reproduction on microfilm or in any other way, and storage in data banks. Duplication of this publication or parts thereof is permitted only under the provisions of the German Copyright Law of September 9, 1965, in its current version, and permission for use must always be obtained from Physica-Verlag. Violations are liable for prosecution under the German Copyright Law.

© Physica-Verlag Heidelberg 1997
Printed in Germany

The use of general descriptive names, registered names, trademarks, etc. in this publication does not imply, even in the absence of a specific statement, that such names are exempt from the relevant protective laws and regulations and therefore free for general use.

Softcover Design: Erich Kirchner, Heidelberg

SPIN 10628282 88/2202-5 4 3 2 1 0 – Printed on acid-free paper

Preface

European environmental and energy policies are currently facing two mutually dependent issues: CO_2 abatement and the completion of the Internal Market for energy. Both of these will trigger substantial structural changes in the energy supply industry and in the wider economy. The purpose of this book is to analyze the interaction between CO_2 abatement, economic structural change and the completion of the European Internal Market. This is a complex task, involving not only significant general equilibrium effects, but also technological changes, especially in the electricity supply industry. Our approach in this study is therefore to combine a multi-region computable general equilibrium framework with a detailed linear-activity representation of the European electricity supply sector.

The project was carried out from September 1994 through August 1996 with the financial support of the Volkswagen-Stiftung. It was jointly undertaken by EWI (Köln) and IER (Stuttgart). However, the authors did not work equally on all chapters of the book. Hoster is primarily responsible for chapter 3 and Böhringer for chapters 4 and 8. Chapters 2, 6 and 7 were jointly written by Hoster and Welsch, and chapters 1 and 5 as well as part III by Hoster, Welsch and Böhringer.

Methodological approaches and partial results have been presented at workshops and seminars in Ladenburg, Berlin, Interlaken, Stuttgart, Paris and Rome.

Several people contributed to or provided advice toward the completion of this project, most noteworthy Ulrich Fahl (IER), Thomas F. Rutherford (University of Colorado) and Hans Fehr (University of Tübingen). Dietrich Hoffmann provided assistance in constructing the database and in completing the manuscript.

Any shortcomings are, of course, our responsibility.

Köln, Stuttgart, F. Hoster, H. Welsch, C. Böhringer

Contents

1 Introduction

1.1 CO_2 Abatement, Structural Change and the European Internal Market

Over the last two decades the environmental impacts resulting from energy generation and utilization have become one of the major concerns of environmental policy and a central issue in environmental and energy economics. The emissions arising from the combustion of fossil fuels are major contributors to the currently most heavily debated environmental issues. In particular, more than half of the contribution to the anthropogenic greenhouse effect is attributed to carbon dioxide (CO_2), whose release is largely related to fossil energy use. Since there are no prospects for economically feasible end-of-pipe technologies for CO_2 abatement, these emissions can only be reduced by scaling back fossil fuel utilization. This creates spillovers to other energy-related pollutants (SO_2, NO_x), because the reduction of fossil energy input implies lower emissions of these pollutants as well. Thus, CO_2 abatement plays a key role in energy-related environmental policy.

In response to the greenhouse problem, several countries have adopted targets to reduce their emissions of CO_2 (see IEA 1994). The European Union aims at stabilizing its emissions by 2000 at the level of 1990. Several member states have even more stringent goals. For example, Germany's official objective is a 25-percent reduction by 2005, relative to the level in 1990. On the other hand, other member states such as Spain have set limits, which permit an increase in their emissions.

Several policy instruments have been suggested as a means of CO_2 abatement. Market-based instruments, such as emission taxes or tradeable emission permits, which have been recommended by economists for decades because of their cost-effectiveness, are now being seriously contemplated even in the political sphere.[1] In 1992 the European Commission submitted a draft directive suggesting the introduction of a combined carbon/energy tax in the European Union. While

[1] One reason for this may be that these instruments are more easily applicable to carbon emissions than to other pollutants, because their measurement does not require sophisticated instrumentation: carbon emissions can be computed from the carbon content of the fuels. In addition, carbon taxes or permits need not be differentiated by spatial origin of emissions, because the origin has no influence on the impact.

efforts to introduce a joint European tax have meanwhile been abandoned, the Council of Environmental Ministers has left it to the member states to introduce such measures on a national basis.

Whether or not CO_2 abatement measures are taken, and whatever form they take, the future development of European CO_2 emissions will be strongly influenced by economic structural change. Conversely, any serious CO_2 abatement policy will have an effect not only on the level, but also on the structure of economic activity.

Three dimensions of structural change have a strong influence on energy-related environmental impacts:

(1) *Changes in the sectoral structure*: Because of sectoral differences in energy intensity and energy mix, changes in the sectoral structure immediately affect the development of emissions. Within sectors, changes in the energy mix have a direct effect on emissions as well. This is especially important with respect to the electricity supply industry, since this industry accounts for almost one third of European CO_2 emissions, with shares in individual countries ranging from 15 percent (France) to about 50 percent (Denmark).

(2) *Technological change*: More efficient energy transformation technologies, such as combined-cycle power stations, allow for the reduction of primary energy input and resulting emissions at any given level of final energy demand. The corresponding emission reduction potentials will, however, only be realized to the degree in which these technologies are economical. This depends, among other things, on policy measures that make primary energy or environmental emissions more costly.

(3) *Institutional change*: An important aspect of institutional change is the organization of markets. In this regard, major impacts are to be expected from the deregulation of European energy markets. Whereas in manufacturing and in services the opening of national markets has largely been completed, corresponding reform of energy markets is still under way. Only recently a timetable has been adopted by the Council of Energy Ministers, according to which national electricity markets will gradually be opened for competition. This has the effect that, to the degree in which international trade in electricity is economical, *national measures* to reduce electricity-related emissions can be evaded by electricity imports. On the other hand, competitive pressure emerging in the wake of deregulation might lead to an accelerated introduction of new, more efficient and, hence, more environmentally compatible generation technologies.

From these considerations it follows that the issues of CO_2 abatement, structural change, and the European Internal Market for energy are intimately intertwined. The interaction of these issues may produce several partial effects of opposing directions, whose overall outcome is difficult to predict unless a general equilibrium perspective is taken.

Our approach in this study is to take changes in the institutional framework, especially energy and environmental policy as exogenous, and to trace their effects, especially with respect to sectoral and technological structure using a general equilibrium framework

The study adopts a European-wide perspective and places special emphasis on the technological foundation of energy substitution possibilities. The European-wide perspective is necessary, because European national economies become increasingly integrated within the European Internal Market and because of the transnational character of energy related pollution. A profound representation of technological options is a prerequisite for the credibility of the quantitative model results.

1.2 Aims and Scope of the Study

The aim of this study is to develop and apply a quantitative analytical tool, which simulates potential impacts of energy policy and environmental policy on the economic structure and emissions for the European Union. Policy issues to be addressed include the following questions:

(1) What would be the effect of the Internal Energy Market on CO_2 emissions, if no specific abatement policies were implemented ?

(2) How would this effect be influenced by national energy policies, especially with respect to nuclear power ?

(3) What would be the (combined) effect on CO_2 of the Internal Energy Market and a European-wide carbon/energy tax ?

(4) What would be the effect on CO_2 of the Internal Energy Market and a carbon/energy tax introduced unilaterally in Germany ?

(5) What would be the economic consequences of attaining CO_2 emission targets through independent domestic reduction policies as compared to coordinated EU-wide abatement?

Out of these issues the first two refer to a situation without specific CO_2 abatement policies. The third and the fourth are concerned with analyzing the effect of applying tax instruments, whereas the fifth considers the use of emission permits to meet given abatement targets.

1.3 Methodological Approach

Given our study aims, a quantitative analytical tool is required that addresses the interaction between market organization, CO_2 abatement policies, and technological dynamics within an integrated and consistent framework.
Several approaches exist for modeling the interaction between energy, the environment and the economy. They differ mainly with respect to the emphasis placed on

(1) a detailed, technology-based treatment of the energy system,

(2) a theoretically consistent description of the general economy.

The models placing emphasis on (1) are typically partial models of the energy sector, lacking interaction with the rest of the economy[2]. In general, they are bottom-up engineering-based linear activity models with a large number of energy technologies. They are mostly used to compute the least-cost method of meeting a given demand for final energy or energy services.

The models emphasizing (2) are general economic models with only rudimentary treatment of the energy system. Following the top-down approach, they describe the energy system (similar to the other sectors) in a highly aggregated way by means of neoclassical production functions, which capture substitution possibilities by means of substitution elasticities. These models may be classified as open (demand driven Keynesian) and closed (general equilibrium) models.

None of these alternatives is appropriate for our purpose. What we aim to develop is a tool with the following features:

(1) A closed, consistent representation of both the demand and the supply side of the economy, i.e., a general equilibrium model. CO_2 abatement is a *long-term* policy issue, whose analysis requires the simultaneous treatment of supply and demand side responses. The general equilibrium model should incorporate a sufficient sectoral and geographical disaggregation of the European economy. A multi-sectoral, multi-country framework is necessary in order to investigate the effects of CO_2 abatement and energy market opening at the sectoral and national level.

(2) A detailed technological representation of the European energy system. With respect to CO_2 emissions the electricity supply industry plays a key role because of : (i) the high share of electricity generation in CO_2 emissions, (ii) the potential for technological change in electricity generation, and (iii) the institutional changes initiated in the European electricity market.

---

[2] One exception is ETA-MACRO (Manne 1981) and its derivatives. It combines a fairly detailed linear technology model of energy supply with a highly aggregated (one sector) macroeconomic model.

This implies that we have to combine an aggregate representation of the general economy based on neoclassical economic theory with a detailed engineering-based representation of individual electricity generation technologies.
There are two ways of achieving this. One is iterative soft-linkage; the other is integrated modeling under the complementarity format.

The first approach uses an equation-based computable general equilibrium model, from which the electricity sector is truncated, and combines it with a detailed linear optimization model of the electricity supply system. The electricity supply model determines the configuration of capacities and of power generation (differentiated by technology and country), which satisfies demand at minimum overall cost, given the prices of inputs (fuel prices, capital costs, emission taxes or permit fees). The resulting input requirements are fed into the general equilibrium model, and both models are iterated until sufficient agreement is reached.

The second approach uses the representation of a (Walrasian) general equilibrium as a so-called complementarity problem to accommodate the hybrid description of production possibilities through inequalities within a single format. This means that those sectors, whose technological options are of major importance for the policy issue (the electricity sector), can be represented through bottom-up activity analysis, whereas the remaining sectors can be described by continuous neoclassical production functions.

Both of these modeling options have their specific merits and shortcomings. The complementarity approach with its single modeling format has advantages in consistency, but large models may cause problems for the robustness of commercially available solvers.[3] Under the iterative approach virtually no limits are placed on the dimensionality of the submodels. Another advantage is the possibility to use existing submodels, which can significantly reduce the efforts of model development. However, different methodologies for defining the submodules can lead to problems in consistency. Decomposition methods employed in the iterative approach are often critical as to the convergence of the partial solutions towards a consistent overall solution.

In this study, both approaches will be employed. The effects of creating an Internal Market for electricity, either as an isolated measure or in combination with various forms of carbon/energy taxation, will be studied using the iterative approach. For the analysis of quantitative restrictions by means of tradeable permits, the complementarity approach will be used. Both approaches will be based on a common classification of regions, sectors and technologies and will use a common data set.

---

[3] See e. g. PATH (Dirkse and Ferris 1995) or MILES (Rutherford 1993).

1.4 Overview

The study is organized in three parts.
Part I provides a description of the models employed for policy analysis and the assumptions underlying our simulations.

Chapter 2 describes the multi-country, multi-sector general equilibrium model LEAN. Items addressed are the mechanism of market clearance and the macro closure, the modeling of foreign trade, final demand, factor demand, prices and taxes and capital formation.

Chapter 3 presents EIREM, a linear activity model of the European electricity generation and transmission system. It states the objective function, the cost equations, restrictions on demand, fuel types, generation and transmission capacities, and environmental constraints (especially environmental taxes).

Chapter 4 provides a detailed description of NEWAGE , a general equilibrium model designed to analyze the economic and environmental impacts of meeting emission abatement targets at both the single country and European level. After a general introduction into the complementarity format the chapter outlines the explicit integration of linear activity analysis into the general equilibrium modeling framework. The final section gives a comprehensive algebraic summary.

Chapter 5 gives an overview of the parametric framework and the policy scenarios. This includes the description of the regional, sectoral and technological classifications, benchmark data and elasticities for the economy-wide models, the characterization of technologies employed in EIREM and NEWAGE, and the choice of exogenous drivers. Subsequently, the basic structural indicators characterizing the initial situation are discussed and the policy scenarios to be studied are defined.

Part II presents the results of the policy simulations performed.

Chapter 6 deals with the economic and environmental consequences of the Internal Energy Market. It describes and interprets the development of environmental and economic indicators in the base case scenario and the changes in this development implied by a European Internal Market for electricity. The analysis is differentiated with respect to the German policy on nuclear power utilization.

Chapter 7 addresses the effects of introducing a carbon/energy tax in the presence of a European Internal Market for electricity. Both a European-wide tax and a national tax solely in Germany are considered. The effect of such taxes on CO_2 emissions, macroeconomic indicators, and the sectoral structure of the various economies is discussed. In comparing the national tax with the European tax, emphasis is placed on the issues of carbon leakage and international competitiveness of the various industries.

In Chapter 8 the effects of achieving declared national emission reduction targets and an equivalent overall target for the EU are analyzed. The policy instruments considered are emission permits, which may be traded either nationally or throughout Europe. For both policy scenarios the resulting abatement costs and the effect on the macroeconomic indicators and the sectoral structure of the various economies are examined.

Part III summarizes and concludes.

Part I: Models and Assumptions

2 LEAN

2.1 Overview

LEAN (Low Emission Assessment eNgine) is a multi-country general equilibrium system to assess the macro-sectoral effects of various carbon dioxide reduction options in the European Union. The time horizon is extended to 2020. The model is recursively dynamic, which permits model solutions for a sequence of temporary equilibria.

For labor and energy, disembodied factor-augmenting technical progress is assumed. For capital, technical progress is of an embodied type, such that the average efficiency of each sector's capital stock can only be increased by introducing new, more modern equipment.

Aggregate labor supply is described by a dynamic wage equation, which explains wage formation through the dynamics of labor productivity in conjunction with a Phillips curve mechanism.

Foreign trade is modelled by means of a world trade pool rather than by modeling bilateral trade relations. In the foreign trade model, the rest of the world is represented by exogenous import volumes and export prices (in terms of the rest of the world's currency).

Both the European Union and the rest of the world are treated as exchange rate unions vis a vis each other. Thus, two currencies may be said to exist in the model. They are linked by a flexible exchange rate, which is assumed to react to changes in the EU's current account vis a vis the rest of the world.

All aggregator functions in the model (quantity and price aggregators) are of the Leontief or CES type. A nested production structure allows the substitution elasticities to differ among sub-aggregates. Consumer demand is modelled by means of the Linear Expenditure System.

The model is calibrated to 1985 input-output tables of the European Union.

The classifications employed in LEAN are summarized in subsection 5.1.1. The model comprises six European regions (Germany, France, United Kingdom, Italy, Belgium/Netherlands, Spain/Portugal) and 13 industries, of which five correspond to energy carriers (hard coal, lignite, oil, gas, electricity). In the model version used in this study, the electricity sector is truncated from the model. Its place is taken by the linear activity model EIREM.

2.2 Market Clearance and Macro Closure

For each good there are several classes of markets: a world market and six domestic goods markets. Also, six domestic labor markets and one international capital market exist.

In the *world market* the world trade volume is determined by summing import demand across the six European regions and the rest of the world (ROW). Since the imports of the European regions are endogenous, a considerable fraction of the world trade volume, to be allocated to exports of the various regions, is endogenous.

Formally we have

$$EX_i^W = \sum_r IM_i^r + IM_i^{ROW}, \tag{2.1}$$

where EX_i^W = world trade volume of good i, IM_i^r = import demand of European region r, IM_i^{ROW} = import demand of the rest of the world.

In the *domestic goods markets* total supply equals the sum of intermediate demand, consumption demand, government demand, investment demand, and export demand:

$$X_i = \sum_{j \in S} XX_{ij} + XC_i + XG_i + XI_i + XEX_i \tag{2.2}$$

where X_i = total supply, XX_{ij} = intermediate demand, XC_i = consumption demand, XI_i = investment demand, XG_i = government demand, XEX_i: = export demand, S = set of sectors.

For the energy sectors, intermediate demand for their own product is split into energy throughput and energy consumption, because only energy consumption leads to emissions of CO_2.[1]

Labor is assumed to be immobile across borders. Employment in each country is determined as being the sum of labor demand across sectors:

$$LAB = \sum_{j \in S} L_j, \tag{2.3}$$

where LAB = employment, L_j = sectoral labor demand.

In contrast to the demand side, which is captured by ordinary demand functions, supply is modeled in an inverse fashion, via supply prices (see section

[1] Energy throughput comprises, e.g., the use of hard coal as 'raw material' for the production of coke.

2.5). Because, in any model solution, the demand functions are evaluated at those supply prices, the overall quantities demanded are equilibrium quantities.[2]

The *capital market* is treated differently. Capital market equilibrium requires that the value of macroeconomic net investment equals private savings less the budget deficit less the balance of current account. This condition is a way of stating the equality of income generated and income used. The way in which a model *achieves* this accounting identity is usually referred to as the macro closure. In the current model we use the interest rate as the closure variable. More specifically, we assume free capital mobility, such that there is a uniform interest rate. Then the *consolidated* flow of funds constraint of the EU vis a vis ROW determines the interest rate:

$$IN - DEPN = s \cdot YVN - (GN - TN) - (EXN - IMN) \rightarrow z \,, \tag{2.4}$$

where IN = investment, $DEPN$ = depreciation, YVN = available income, GN = government expenditure, TN = total tax revenue, EXN = exports, IMN = imports, s = savings ratio, z = real interest rate. (Symbols containing 'N' refer to nominal variables.)

As the *numeraire* we choose the average consumer price level in the six European regions:

$$\sum_r \alpha^r PC^r \equiv 1 \,, \tag{2.5}$$

where PC^r = consumer price index in region r, α^r = share of region r's GDP in total GDP of the European regions.

2.3 Foreign Trade and Final Demand

As is common in computable general equilibrium models, foreign trade modeling follows the approach of Armington (1969), according to which imported and domestically produced goods of the same kind are treated as incomplete substitutes. Thus, the aggregate amount of each good is divided among the two supply sources, *imports* and *domestic production*:

$$IM_i = f(X_i, PX_i, PIM_i) \,, \tag{2.6}$$

$$Q_i = f(X_i, PX_i, PQ_i) \,, \tag{2.7}$$

---

2 This is a convenient and frequently-employed method of computing equilibria (see, e.g. Conrad 1993).

where PIM_i = import price, IM_i = import volume, PX_i = price of total supply, Q_i = domestic output, PQ_i = price of domestic output.

For *exports*, there is a similar, but nested, structure. First, the world trade volume of each good is allocated to exports from ROW and exports from the EU. The latter is then subdivided into exports from the various EU regions:

$$XEX_i^{EU} = f(EX_i^W, PEX_i^W, PEX_i^{EU}) , \tag{2.8}$$

$$XEX_i^r = f(XEX_i^{EU}, PEX_i^{EU}, PEX_i^r) , \tag{2.9}$$

where PEX_i^W, PEX_i^{EU}, PEX_i^r denote the export prices of the world, the EU, and of region *r*.

The incomplete substitutability between goods of different origin is captured by CES aggregator functions. Accordingly, the demand for imports and domestic production, as well as the two levels of export demand, are determined by CES demand functions.[3]

Turning to *consumption*, we assume that the consumption expenditures of the representative household are a fraction of available labor and capital income. The savings ratio is assumed to depend on the interest rate, with a constant elasticity.[4] Total consumption expenditures are then allocated to the consumption of different goods, utilizing the Linear Expenditure System:

$$CN = (1-s) \cdot YVN = (1-s) \cdot (WAGE + NOS) , \tag{2.10}$$

$$s = a \cdot z^b , \tag{2.11}$$

$$XC_i = \overline{XC}_i + \gamma_i \frac{CN - \sum_j \cdot PX_j \cdot \overline{XC}_j}{PX_i} , \tag{2.12}$$

where CN = consumption expenditures, $WAGE$ = labor income, NOS = net operating surplus, (a, b, γ_i: parameters).

Government expenditure in nominal terms is determined as a constant fraction of nominal GDP in the previous period, and real government expenditure is obtained using this result and dividing by the price of the sector non-market services[5]:

---

3 CES demand functions (and price functions) are compiled, in general terms, in section 2.7.

4 This encompasses the frequently-considered special case of a constant savings ratio.

5 In the input-output table the column 'government expenditure' is non-empty only in the row 'non-market sevices'.

$$GN = s_{GY} \cdot YN_{-1} \ , \tag{2.13}$$

$$XG = GN \ / \ PX_{NMSV} \ , \tag{2.14}$$

where YN =nominal GDP, PX_{NMSV} = price of non-market services (s_{GY}: parameter).

Nominal macroeconomic *investment* is the sum of the sectors' investment in value terms. The price of investment goods is sector specific, since each sector's capital good is characterized by its specific composition in terms of sectors of origin. A sector's real investment is the difference between the capital stock considered optimal for the next period (see section 2.6) and that part of the current capital stock that will still be in operation in the next period. Finally, investment demand for a sector's goods is the sum of sectoral investment requirements, weighted by the (constant) coefficients of the capital composition matrix:

$$IN = \sum_{j \in S} PI_j \cdot I_j \ , \tag{2.15}$$

$$I_j = K_j^* - (1 - \delta_j) K_j \ , \tag{2.16}$$

$$XI_i = \sum_{j \in S} \kappa_{ij} \cdot I_j \ , \tag{2.17}$$

where I_j = sectoral investment, PI_j = sectoral purchase price of capital, K_j = capital stock at beginning of period, K_j* = capital stock planned for next period, δ_j = depreciation rate, κ_{ij} =coefficients of capital composition matrix.

2.4 Factor Demand

Factor demand is derived from a five-stage nested production function for each sector allowing for a flexible treatment of substitution possibilities. *Figure 2.1* displays the production hierarchy.

Figure 2.1. Production Hierarchy

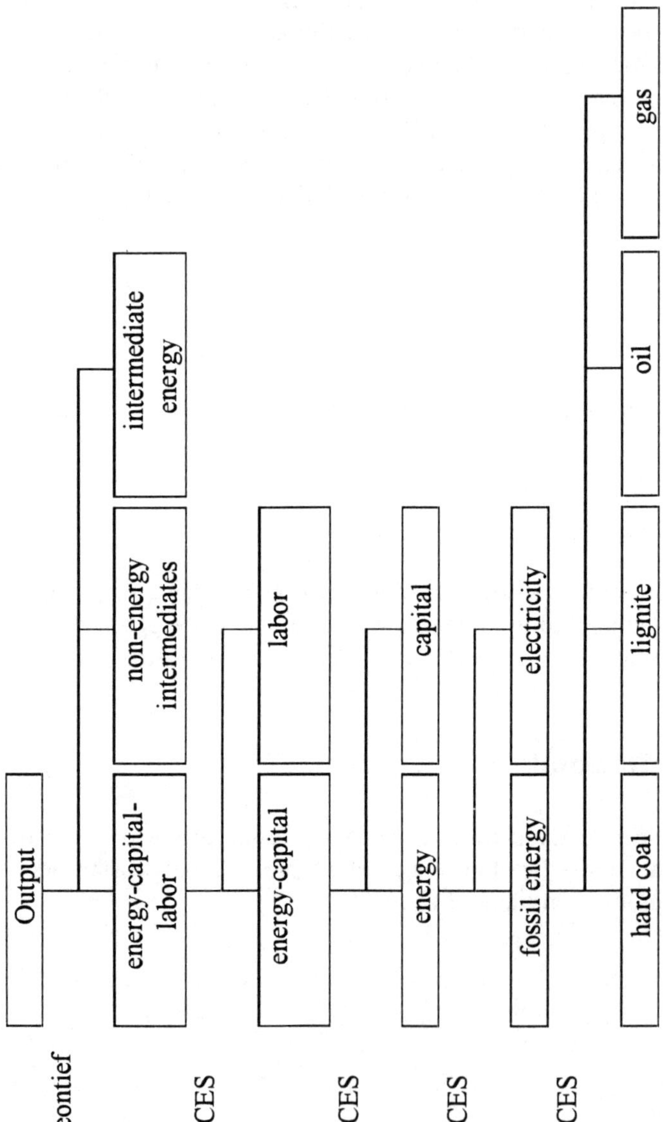

At the top level, output is linked to an *aggregate of energy, capital and labor* (*EKL*) and to the various non-energy *intermediate inputs* via constant input-output coefficients. Note that energy in the *EKL* aggregate is to be understood in the narrow sense, i.e., it comprises only the 'energetic' use of energy carriers. The 'non-energetic' use is separated out and treated as an intermediate input. Non-energetic use is considered for the energy sectors only, where it is taken to be a constant fraction of output.

The *EKL* aggregate is broken down further into *labor* and an *energy-capital aggregate*. This choice of disaggregation reflects our interpretation of capital as a collection of facilities for using energy. While the substitution possibilities *within* the energy-capital aggregate are small, they are larger *between* energy-capital and labor.[6]

Next, energy-capital is separated into capital and energy. Energy, in turn, is an aggregate of fossil energy and electricity. Finally, fossil energy is composed of the four different fossil fuels distinguished in the model.

Typically, the elasticity of substitution among fossil fuels is larger than that between fossil fuels and electricity. The latter, in turn, is larger than that between energy and capital.[7]

Factor demand is derived from profit maximization subject to the production structure just outlined. At the top level of the production process inputs are related to output via fixed coefficients. At subsequent levels CES demand functions are similar to those used in the foreign trade model (for the precise form, see section 2.7). The sectoral capital stock (in effective units) in operation in any period is the capital stock considered optimal in the previous period.[8] Thus, capital is a quasi-fixed factor. Energy, being a variable factor, adjusts to the predetermined capital stock.

Formally, the hierarchy of factor demands is described as follows:

$$XX_{ij} = f(Q_j) , \tag{2.18}$$

$$EKL_j = f(Q_j) , \tag{2.19}$$

$$\tilde{L}_j = f(EKL_j, PEKL_j, \tilde{P\tilde{L}}) , \tag{2.20}$$

$$EK_j = f(EKL_j, PEKL_j, PEK_j) , \tag{2.21}$$

---

6 For an extensive discussion in conceptual and empirical terms see Burniaux et al. (1992).

7 For an overview of estimates see again Burniaux et al. (1992).The assumptions made in the current study are given in subsection 5.1.2.

8 For a derivation of the optimal capital stock see section 2.6.

$$\tilde{K}_j = \tilde{K}^*_{j,-1} , \qquad (2.22)$$

$$\tilde{E}_j = f(EK_j, \tilde{K}_j) , \qquad (2.23)$$

$$XXEC_{EL,j} = f(E_j, PE_j, PEL_j) , \qquad (2.24)$$

$$XXEC_{F,j} = f(E_j, PE_j, PF_j) , \qquad (2.25)$$

$$XXEC_{F1,j} = f(XXEC_{F,j}, PF_j, PF1_j) , \qquad (2.26a)$$

$$\vdots$$

$$XXEC_{F4,j} = f(XXEC_{F,j}, PF_j, PF4_j) , \qquad (2.26d)$$

where the notation is as follows: EKL_j =energy-capital-labor aggregate, L_j = labor, $PEKL_j$ = price of energy-capital-labor aggregate, PL = wage rate, EK_j = energy-capital aggregate, PEK = price of energy-capital aggregate, K_j = capital stock, E_j = energy aggregate, $XXEC_{EL,j}$ = electricity input, PE_j = price of energy aggregate, PEL_j = price of electricity, $XXEC_{F,j}$ = fossil fuel aggregate, PF_j = price of fossil fuel aggregate, $XXEC_{F1,j}$..., $XXEC_{F4,j}$ = individual fossil fuel inputs, $PF1$,..., $PF4$ = price of individual fossil fuel inputs. Tildas refer to variables in 'effective units'.

The theoretical derivation of the planned capital stock is presented in section 2.6.

2.5 Prices and Taxes

Prices represent the supply side of the model. The exogenous driving forces of the price model are the export prices of the rest of the world, expressed in the ROW currency, and the carbon and energy taxes. The model provides the possibility to tax the various production sectors and the households at different rates.

Due to the assumption of perfect intersectoral mobility of labor, a uniform *wage rate* in each region is inherent. The current wage equals the wage of the previous period times the increase in labor productivity and the price level change, modified by the ratio between actual employment and 'normal' employment. This may be seen as a dynamic version of the Phillips curve. This wage equation is equivalent to a labor supply function, according to which labor in excess of its normal level is attracted, whenever wages increase by more than the growth rate of productivity:

$$PL = \Pi \cdot PL_{-1} \cdot \left(\frac{LAB}{\overline{\overline{LAB}}} \right)^{\alpha} \qquad \left(\Leftrightarrow LAB = \overline{LAB} \cdot \left(\frac{PL}{\Pi \cdot PL_{-1}} \right)^{1/\alpha} \right) \tag{2.27}$$

where Π = rate of nominal productivity growth.

As mentioned in section 2.2, the uniform *interest rate* is determined by the requirement of capital market equilibrium.

For the *exchange rate* of the EU vis a vis ROW (price of ROW currency), we specify a dependence on the consolidated balance of the EU's current account. More specifically, we assume that the exchange rate remains constant, whenever the current account balance is in equilibrium. A negative balance represents excess demand for foreign exchange and induces an increase in the exchange rate, while a positive balance works in the opposite direction:

$$ER = ER_{-1} \cdot \left(\frac{\sum_r IMN_r}{\sum_r EXN_r} \right)^{\beta} \quad , r \in \{GER, EC9\} , \tag{2.28}$$

where ER = exchange rate EU vs. ROW.

The prices and taxes discussed so far represent the 'fundamentals' of the price model. All other prices are derivatives thereof. These derivatives are obtained using *price aggregator functions* dual to the quantity aggregators (production functions) discussed above. These price aggregator functions may be interpreted as marginal cost functions, or inverted supply functions. The assumption of constant returns to scale, implicit in the CES specification of the production functions, implies constant marginal costs, which are at the same time average costs. Hence there are no quantities among the arguments of the price functions[9]. Only for the price of the energy-capital aggregate is a dependence on the level of the aggregate and on the capital stock apparent. This is due to capital's fixed nature (in the short term), which implies that marginal and average costs differ and neither of them is constant.

Formally, the price aggregates are as follows:

$$PEX_i^W = f(PEX_i^{ROW} \cdot ER, PEX_i^{EU}) \equiv PIM_i^r , \tag{2.29}$$

$$PEX_i^{EU} = f(PX_i^{GER}, PX_i^r) , \tag{2.30}$$

$$PX_i = f(PQ_i, PIM_i) , \tag{2.31}$$

$$PQ_i = f(PX_i, PEKL_i) , \tag{2.32}$$

[9] The functional form is stated in section 2.7.

$$PEKL_i = f(PL, PEK_i) \ , \tag{2.33}$$

$$PEK_i = f(EK_i, K_i, PE_i) \ , \tag{2.34}$$

$$PE_i = (PEL_i, PF_i) \ , \tag{2.35}$$

$$PF_i = f(PF1_i, \ldots, PF4_i) \ , \tag{2.36}$$

$$\left.\begin{matrix} PEL_i \\ PF1_i \\ \vdots \\ PF4_i \end{matrix}\right\} = PX_j + (1 - coef_{ji}) \cdot \left[c_j e_j TC_{ji} + e_j TE_{ji}\right] \tag{2.37}$$

where TC_{ji} = carbon tax rate (ECU/tCO_2), TE_{ji} = energy tax rate (ECU/TOE), c_j = specific carbon content of energy; e_j = specific energy content of energy good; $coef_{ji}$ = share of non-energetic use of energy good j in sector i.

Of special interest is the world market price of the different goods, as given in eq. (2.29). The world market price is an aggregate of the export prices of the EU and ROW and is expressed in terms of the EU currency. Thus, it depends on the exchange rate. It is identical to the import price of the EU countries.

Also of interest is the price of electricity and the price of the fossil fuels, given in eq. (2.37). The purchase price of these energy carriers equals their generation price plus the carbon and energy tax. Note that these energy prices may differ by user, due to possibly varying tax rates. Also observe that only energy *consumption* is subject to taxation, not the use of energy carriers for non-energy purposes.

2.6 Capital Formation

Our theory of capital formation differs from the usual approach by taking account of embodied technical progress. The theory is meant to apply to each sector separately, but sector indices are omitted in the presentation.

The usual stock-flow accounting relationship for capital is:

$$K_t = (1 - \delta)K_{t-1} + I_{t-1} \tag{2.38}$$

Observe that investment becomes effective with a one-period delay.

Because there is capital-augmenting technical progress, it is useful to introduce the concept of capital efficiency. Our key assumption on capital efficiency is that the efficiency of the existing capital stock (a_t) involves a weighted average of the efficiency of last period's capital stock (a_{t-1}) and the efficiency of the latest vintage now in operation (b_{t-1}):

$$a_t = \frac{(1-\delta)\,K_{t-1}}{K_t}\,a_{t-1} + \frac{I_{t-1}}{K_t}\,b_{t-1} \tag{2.39}$$

The efficiency of the latest vintage is assumed to grow at an exogenous rate.

This equation can be rearranged to give the stock-flow accounting relationship for capital in efficiency units:

$$\begin{aligned} a_t\,K_t &= (1-\delta)\,a_{t-1}\,K_{t-1} + b_{t-1}\,I_{t-1} \\ \tilde{K}_t &= (1-\delta)\,\tilde{K}_{t-1} + \tilde{I}_{t-1} \end{aligned} \tag{2.40}$$

By turning to continuous time, the following equation of motion of the capital stock in efficiency units is obtained:

$$\dot{\tilde{K}}_t \approx \tilde{K}_t - \tilde{K}_{t-1} = \tilde{I}_{t-1} - \delta\,\tilde{K}_{t-1} \tag{2.41}$$

Using the variable cost function[10]

$$VC = VC(\underset{+}{EK}, \underset{-}{\tilde{K}}, \underset{+}{\tilde{P}\tilde{E}}) \tag{2.42}$$

the intertemporal cost minimization problem can now be stated as:

$$\begin{aligned} &\min_{\tilde{I}} \int_0^\infty e^{-zt}\left[VC(EK, \tilde{K}, \tilde{P}\tilde{E}) + \tilde{P}\tilde{I}\cdot\tilde{I}\right]dt \\ &s.t.\ \dot{\tilde{K}} = \tilde{I} - \delta\,\tilde{K}, \end{aligned} \tag{2.43}$$

where $\tilde{P}\tilde{I} = PI\,/\,b$.

With static price expectations the optimal investment program is then characterized by the condition that a marginal decrease in variable costs brought about by expanding the capital stock (in efficiency units) should just match the user costs incurred ($\tilde{u}\tilde{c}$):

[10] The variable cost function gives the minimum costs to be spent on variable factors, conditional on variable factor prices and the quantities of output and of the fixed factor.

$$-VC_K = \tilde{u}\tilde{c} \overset{CES}{\Leftrightarrow} \tilde{K}^* = \left[\left(\frac{\tilde{u}\tilde{c}}{\widetilde{PE}} \frac{d_E}{d_K}^{1/r}\right)\frac{r}{1-r} + d_K\right]^{-1/r} EK^e \tag{2.44}$$

where $\tilde{u}\tilde{c} = (z + \delta + w_b) \cdot \widetilde{PI}$.

Note that, in contrast to the usual case, the user cost expression includes the growth rate of the efficiency of new capital (w_b). This indicates opportunity costs in terms of technical progress foregone by investing 'now', rather than 'later'.

In the case of CES production functions and in the absence of adjustment costs, this first-order condition can be readily solved for the optimal capital stock in efficiency units. Because capital formation always takes place one period in advance, expectations as to the activity level need to be formed. We assume that the expected activity level governing capital formation (EK^e) is obtained by extrapolating past growth rates, i.e. myopic expectations with respect to growth rates are formed.

2.7 Functional Forms

The quantity aggregator functions in the foreign trade model are specified as CES functions. The production functions are of a nested Leontief/CES form.

Maximizing profit or utility on the basis of these quantity aggregators yields demand functions, which, in turn, can be used to compute cost functions[11]. Our special interest lies in average cost functions, which we interpret as inverse supply functions. To emphasize the duality between quantities and prices, average cost functions will be referred to as price aggregators.

Consider the following CES quantity aggregator function (production function)

$$X = \left[\sum_{i=1}^{n} d_i (g_i X_i)^{\frac{\sigma-1}{\sigma}}\right]^{\frac{\sigma}{\sigma-1}} \tag{2.45}$$

where X_i= input of good i in quantity units,
$g_i X_i$= input of good i in effective units,
σ= elasticity of substitution,
d_i= distribution parameter.

---

[11] Frequently, cost functions are taken as the basic description of technology. Demand functions are then obtained utilizing Shephard's lemma. Our exposition arbitrarily starts from production functions as the basic concept.

If $g_i \geq 1$ increases over time, this is referred to as factor-augmenting technical progress. If, conversely, $g_i=1$; the technology is stationary.

If all inputs are variable, the demand functions corresponding to (2.43) are

$$X_i = d_i^{\sigma} \left(\frac{PX}{PX_i} \right)^{\sigma} g_i^{\sigma - 1} X , \qquad (2.46)$$

and the price aggregator is of the form

$$PX = \left[\sum_i d_i^{\sigma} \left(\frac{PX_i}{g_i} \right)^{1-\sigma} \right]^{\frac{1}{1-\sigma}} \qquad (2.47)$$

These formulas are valid for the foreign trade model and the production model, except for the top level and the *EK* level.

At the top level, demand is obtained from fixed input-output coefficients, and the price aggregator is the weighted average of input prices.

At the *EK* level, there are only two inputs to the quantity aggregator (2.43), one of which (capital) is fixed. The demand for the other factor (energy) is then obtained by rearranging the eq. (2.43). The profit-maximizing supply price has the form

$$PEK = \frac{\widetilde{P}\widetilde{E} \cdot \widetilde{E}}{EK - d_K EK^{\frac{1}{\sigma}} K^{\frac{\sigma-1}{\sigma}}} , \qquad (2.48)$$

where the variables indicated by a tilda are in effective units.

To see this, consider the profit-maximization problem at the *EK* level:

$$\max_{\widetilde{E}} \; PEK \cdot EK - \widetilde{P}\widetilde{E} \cdot \widetilde{E} - \widetilde{u}\widetilde{c} \cdot \widetilde{K} , \qquad (2.49)$$

in which $\widetilde{K}$ is predetermined. This gives the first-order condition

$$\frac{\partial \, EK}{\partial \, \widetilde{E}} = \frac{\widetilde{P}\widetilde{E}}{PEK} . \qquad (2.50)$$

Because the *EK* aggregator function is linear homogenous, we have

$$EK = \frac{\partial EK}{\partial \tilde{E}} \tilde{E} + \frac{\partial EK}{\partial \tilde{K}} \tilde{K}. \tag{2.51}$$

Inserting (2.48) into (2.49) and solving for *PEK* gives:

$$PEK = \frac{\tilde{P}\tilde{E} \cdot \tilde{E}}{EK - \frac{\partial EK}{\partial \tilde{K}} \tilde{K}}. \tag{2.52}$$

Finally, when observing the functional form of the *EK* aggregator we obtain

$$\frac{\partial EK}{\partial \tilde{K}} = d_K \left(\frac{EK}{\tilde{K}} \right)^{1/\sigma}, \tag{2.53}$$

which together with (2.52) gives (2.48).

3 EIREM

3.1 Overview

EIREM (European Inter-Regional Electricity Model) is a dynamic linear-optimization model of the European power generation and transmission system.[1] It operates in five-year steps from 1985 through 2055. The extended time horizon is chosen to avoid the so-called 'end effect', in which repair of existing capacity is substituted for capacity additions near the end of the planning period. The actual evaluation of results extends to 2020.

The objective function is the sum of total discounted generation and transmission costs, which is minimized subject to a given demand profile and, in the case of closed national markets, constraints on maximum allowed imports. The decision variables are the utilization rates of the capacities existing in each period (short term) and the addition and removal of capacities (long term), where 'capacities' refers to both generation and transmission.

Demand is differentiated by region, year, season, and load period. The model distinguishes three seasons (summer, interseasonal time, winter) and four load periods (high, medium, low, and base load).

Existing and planned generation capacities are differentiated by plant type (technology) and vintage (expressed in terms of five-year intervals). Each vintage of each technology is characterized in terms of specific investment costs and heat rates. Heat rates together with the carbon content of the various fuels determine the resulting CO_2 emissions. More modern plants have a higher heat rate (less fuel requirement). Since there is no fixed plant lifetime, decommissionings are based on optimization. With increasing operation costs, due to, for instance, emission taxes, inefficient plants will be replaced earlier.

EIREM comprises Germany, France, Italy, the United Kingdom, Belgium/Netherlands and Spain/Portugal as so-called core regions. In addition to these countries the Nordic and the East European countries are included as so-called satellite regions. Their electricity imports are given exogenously, while their exports into the core regions are price-dependent.

---

1 The following exposition is based on Hoster (1996a).

3.2 Objective Function

EIREM is a dynamic model involving the optimization of power supply in central Europe. The objective function to be minimized is the total discounted cost within the core regions r. Total costs are the sum of total operating costs (*TOC*), total labor costs (*TLC*), maintenance and repair costs (*TIURC*), capital costs (*TCC*), expenses for electricity imports from satellite regions (*TIMC*), cost of transmission capacity additions (*TTC*), as well as potential payments due to a CO_2-emission tax (*ECTAX, TOEPFTAX*).

$$TDC = \sum_{r}\sum_{t} \delta_t \cdot (TOC_{t,r} + TLC_{t,r} + TIURC_{t,r} + TCC_{t,r} + TIMC_{t,r} + ECTAX_{t,r} + TOEPFTAX_{t,r}) + \sum_{t} \delta_t \cdot TTC_t \quad , \tag{3.1}$$

where the parameter δ_t denotes the discount factor in period t.

3.3 Cost Equations

Total operating cost comprises fuel costs and all other costs that depend on actual power generation:

$$TOC_{t,r} = \sum_{j}\sum_{v \le t}\sum_{s}\sum_{bp} \frac{PRODUC_{j,v,s,bp,t,r} \cdot (dur_{s,bp} + startdur_{j,s,bp})}{eta_{j,v,r}} \cdot pfuel_{j,t,r} + \sum_{j}\sum_{v \le t}\sum_{s}\sum_{bp} PRODUC_{j,v,s,bp,t,r} \cdot (dur_{s,bp} + startdur_{j,s,bp}) \cdot ovarc_{j,t,r} \quad , \tag{3.2}$$

t = 1985,..., 2055;
r = Germany, France, Italy, U.K., Belgium/Netherlands, Spain/Portugal.

Fuel prices and other specific variable costs enter the model through the exogenous parameters *pfuel* and *ovarc*. Total fuel requirement of a plant within a certain period of time can be computed from the electricity generated and the plant's efficiency (*eta*). Total fuel requirement of the system in region r in period t is the sum of the seasonal (index s) fuel consumption of all power plant types j with vintage $v \le t$. Work is the product of power and the duration of its employment.

The variable *PRODUC* measures power output of plant j with vintage v in season s and mode of operation bp in year t and region r. The index bp denotes the mode in which a plant operates. With four load-categories b there are also four modes in which a power plant can operate. A power plant operating in the largest mode supplies power continuously over all load-categories b. This mode is called the 'base-load' mode. The subsequent mode decreases by the duration of the previous load block. The modeling of load categories is illustrated in *Figure 3.1*.

The plant's time of operation in mode bp considers the duration of the mode of operation (*dur*) and the time needed to bring a plant unit into operation (*startdur*). Generally speaking, however, the latter can be neglected in a yearly analysis.

Specific maintenance and repair costs (*iurcost*) of plant j are determined by its age. Total maintenance and repair costs of a region are summed up for all plant types and vintages:

$$TIURC_{t,r} = \sum_{j} \sum_{v \leq t} iur\cos t_{j,v,t,r} \cdot CAPA_{j,v,t,r} \, . \tag{3.3}$$

Total labor costs are a function of installed capacity as well:

$$TLC_{t,r} = \sum_{j} \sum_{v \leq t} lab\cos t_{j,v,t,r} \cdot CAPA_{j,v,t,r} \, . \tag{3.4}$$

Figure 3.1. Load-Categories

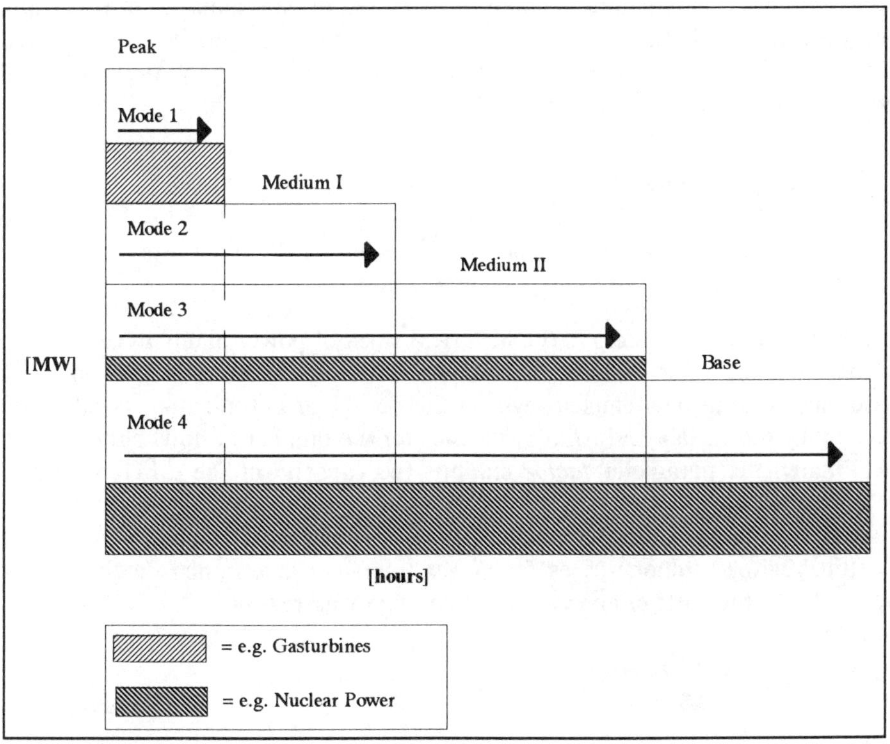

Additions to transmission capacity between two regions show regional differences in specific costs per MW. These specific costs per MW (*tc*) depend on the kind of transmission line (*tl*), on the year of addition (*tn*) and on the regions involved. The kind of addition is determined by the capacity situation of both regional lines and back-to-back stations. According to the situation, more or less costly measures have to be taken to upgrade transmission capacity. The variable *TCADD* stands for transmission capacity of kind *tl* built in year *tn* connecting core regions (*r,reg*) and/or satellite regions (*satr*):

$$\begin{aligned} TTC_t = & \sum_r \sum_{reg} \sum_{tn \le t} \sum_{tl} tc_{tl,tn,t,r,reg} \cdot TCADD_{tl,tn,r,reg} \\ & + \sum_r \sum_{satr} \sum_{tn \le t} \sum_{tl} tc_{tl,tn,t,r,satr} \cdot TCADD_{tl,tn,r,satr} \end{aligned} \tag{3.5}$$

Total capital costs are determined by the cost of investment in new capacity as well as the fixed costs of existing plants. They are defined as fixed, since no decision in the future will influence them. Fixed costs of existing plants must be taken into account to get a clear picture of the development of costs through time. Otherwise total costs would be too low in initial years compared to later periods. Capital costs include the costs of desulphurization and denitrification of flue gases according to national law, as well. After the plant has been written-off, which usually occurs before the plant is taken out of operation, no further capital costs are incurred. Capital costs are accounted for through their annuity (*cc*), which depends on the investment cost of plant *j* with vintage *v*, the interest rate and the age of the power station:

$$\begin{aligned} TCC_{t,r} = & \sum_j \sum_{tp} cc_{j,tp,t,r} \cdot inicap_{j,tp,r} \\ & + \sum_j \sum_{tn \le t} cc_{j,tn,t,r} \cdot factor_{j,tn,tn,r} \cdot NUM_{j,tn,tn,r} \end{aligned} \tag{3.6}$$

The matrix *inicap* refers to the historical stock of power plants in the baseyear of optimization. The index *tp* comprises the sum of all vintages up to the baseyear. Adding new capacity within the period of optimization is taken into account by the variable *NUM*. *NUM* stands for the number of units built in period *tn*. Finally, the parameter *factor* exhibits the capacity of the different units in MW.

Finally, total cost must contain expenses for electricity imports (*PIMP*) from satellite regions. Import prices for power (*pim*) are determined on the basis of avoided variable costs of generation in the importing region.

$$TIMC_{t,r} = \sum_s \sum_b \sum_{satr} PIMP_{s,b,t,satr,r} \cdot duration_{b,s} \cdot pim_{s,b,t,satr,r} \tag{3.7}$$

3.4 Demand Restrictions

The power demand of region r is either met by the power supply within the region or by net power imports:

$$\sum_{bp\geq b}\sum_{j}\sum_{v\leq t} PRODUC_{j,v,s,bp,t,r}\cdot e_{r,r} + \sum_{reg\neq r}\left(PT_{s,b,t,reg,r} - \frac{PT_{s,b,t,r,reg}}{e_{r,reg}}\right) + \sum_{reg\neq r}\left(shterm_{s,b,t,reg,r} - shterm_{s,b,t,r,reg}\right) + \sum_{satr}\left(PIMP_{s,b,t,satr,r} - \frac{PEXP_{s,b,t,r,satr}}{e_{r,satr}}\right) \geq d_{s,b,t,r}, \qquad (3.8)$$

t = 1985,..., 2055;
s = summer, interseasonal time, winter;
b = high, medium, low, base;
r = Germany, France, Italy, U.K., Belgium/Netherlands, Spain/Portugal.

The variable *PT* measures the net power transmitted between the core regions caused by systematic cost advantages. Due to transmission losses, it is necessary to include exports as gross power transmissions. In the EIREM-model short-term optimization of power transfers between regions are not endogenously determined. Such short-term transfers arise, when power plants experience temporary outages or when surprise changes in demand occur. Using historical transfers as a benchmark, the parameter *shterm* has been incorporated to account for these short-term fluctuations and is fitted proportionally to the development of power demand in the regions in question. Additionally, it is assumed that such transfers observe a fixed relationship to the regions´grid load.

The parameter e measures the efficiency of the transmission link between two regions and its value is strictly less than unity.

Domestic power generation is also scaled by a factor e, which corresponds to transmission losses within region r. *PIMP* stands for the power imports from the satellite regions and *PEXP* represents the power exports to them.

Transmission links exist between the various regions. The modeling of the core-regions assumes fixed locations of representative power consumers (C) and producers (P). These locations determine the length of transmission lines. The power demand of the satellite regions (power exports of the core regions) is given exogenously, and they sell power at an exogenously determined price. The parameter $d_{s,b,t,r}$ denotes the demand for electricity (MW) in season s of period t, load characteristic b and region r.

The available installed capacity and the difference between net power imported and gross power exported has to meet the expected peak demand plus a certain reserve margin. In eq. (3.9) it is assumed that the peak demand will occur during the winter season (s). In the program a subroutine identifies the period with the peak demand automatically. A regional reserve margin is required to maintain security of supply. Availability depends on the power plant's technical specifications, on the season and on the age of the plant. While seasonal differences in availability are caused by maintenance interruptions, which mainly occur in the summer season, increasing age reduces availability due to the need for more intensive maintenance and repair.

$$
\begin{aligned}
&e_{r,r} \cdot \sum_{j}\sum_{v \le t} avail_{j,v,s,t,r} \cdot CAPA_{j,v,t,r} + \sum_{reg \ne r}\left(PT_{s,b,t,reg,r} - \frac{PT_{s,b,t,r,reg}}{e_{r,reg}}\right) \\
&+ \sum_{reg \ne r}\left(shterm_{s,b,t,reg,r} - shterm_{s,b,t,r,reg}\right) + \sum_{satr}\left(PIMP_{s,b,t,satr,r} - \frac{PEXP_{s,b,t,r,satr}}{e_{r,satr}}\right) \ge \left(1 + prr_r\right) \cdot d_{s,b,t,r}
\end{aligned}
\quad ,(3.9)
$$

j = HC,..., HYD_P;
v =1955,..., t;
t = 1985,..., 2055;
s = winter;
b = high.

3.5 Capacity Restrictions

The capacity of plant j with vintage v depends on the initial capacity *inicap* and capacity adjustments in subsequent years *tn*. Capacity adjustment can either mean an addition to or replacement of existing plants. The variable $NUM_{j,v,tn,r}$ accounts for the number of plants with size „$factor_{j,v,tn,r}$" that will be added or replaced.

$$
CAPA_{j,v,t,r} = inicap_{j,v,r} + \sum_{tn \le t} factor_{j,v,tn,r} \cdot NUM_{j,v,tn,r} \quad . \qquad (3.10)
$$

j =HC,..., HYD_P;
v = 1955,..., t;
t = 1985,..., 2055.

The output from each plant cannot exceed its available capacity:

$$
\sum_{bp} PRODUC_{j,v,s,bp,t,r} \le avail_{j,v,s,t,r} \cdot CAPA_{j,v,t,r} , \qquad (3.11)
$$

j =HC,..., HYD_P;
v = 1955,..., t;
s = summer, winter, interseasonal time;
t = 1985,...,2055.

3.6 Fuel Input and Generation Restrictions

Restrictions in fuel input are either due to political standards or due to natural limitations. Lignite, for example, can only be found in Germany. The amount of lignite (*BC*) used in generation, can be calculated from the electricity produced divided by the plant's efficiency (*eta*).

$$\sum_{v \le t} \sum_{s} \sum_{bp} \frac{PRODUC_{j,v,s,bp,t,r} \cdot (dur_{s,bp} + startdur_{j,s,bp})}{eta_{j,v,r}} = BC_{t,r}, \tag{3.12}$$

$t = 1985,...,2055;$
$j = BC.$

Power production with lignite can be varied only to a small extent. The amount of lignite used for generation is therefore limited by an upper and a lower bound. Similarly, input of other fossil fuels may be restricted. These limitations vary with the different political scenarios to be analyzed. A different kind of restriction concerns hydroelectric capacity, as the output from a hydroelectric plant cannot exceed the energy available in the water supplies. Especially pumped storage plants have a very low average load factor:

$$\sum_{v \le t} \sum_{s} \sum_{bp} PRODUC_{j,v,s,bp,t,r} \cdot dur_{s,bp} \le lfhyd\_P_r \cdot TOTCAP_{j,t,r}, \tag{3.13}$$

$j =$ HYD_P.

The parameter *lfhyd_P* represents the plant's load factor, which is the quotient of the average yearly production divided by maximum production (see Turvey and Anderson 1977, p. 187). The variable *TOTCAP* accounts for the total capacity of a certain technology.

3.7 Interregional Transmission Constraints

In every season, load, period and region the sum of power flows in opposite directions is restricted by the sum of the initial transmission capacity (*initrans*) and subsequent capacity additions (*TCADD*). The parameter *griduse* restricts maximum transmission capacity to an economically effective level. It must be sufficient to accommodate gross power flows. Comparative cost differences among the regions and transmission losses prevent the model from transmitting power in opposite directions during the same load-period. Hence, nonlinearities due to efficiency improvements in the case of simultaneous transmissions will not occur.

$$\frac{PT_{s,b,t,r,reg} + PT_{s,b,t,reg,r}}{e_{r,reg}} + \frac{shterm_{s,b,t,r,reg} + shterm_{s,b,t,reg,r}}{e_{r,reg}} \leq$$
$$griduse_{r,reg} \cdot \left(initrans_{r,reg} + \sum_{tn \leq t} \sum_{tl} TCADD_{tl,tn,r,reg} \right), \qquad (3.14)$$

s = summer, winter, interseasonal time;
t = 1985,..., 2055;
b = high, medium, low, base;
r, reg = Germany, France, Italy, U.K., Belgium/Netherlands, Spain/Portugal.

$$\frac{PIMP_{s,b,t,satr,r} + PEXP_{s,b,t,r,satr}}{e_{r,satr}} \leq$$
$$griduse_{r,satr} \cdot \left(initrans_{r,satr} + \sum_{tn \leq t} \sum_{tl} TCADD_{tl,tn,r,satr} \right), \qquad (3.15)$$

s = summer, winter, interseasonal time;
t = 1985,..., 2055;
b = high, medium, low, base;
r = Germany, France, Italy, U.K., Belgium/Netherlands, Spain/Portugal;
$satr$ = Scandinavia, Eastern Europe.

Concerning the extension of cross-frontier transmission capacity between two regions three different steps are distinguished. The first and, in terms of costs, cheapest step refers to the upgrading of back-to-back stations. The parameter *tl1up* marks the upper limit of this measure. Additional transmission capacity beyond the improvement of back-to-back stations can be obtained by improving parts of national transmission lines, which is limited by *tl2up*. If these measures still are not sufficient, extensive sections of national lines will have to be upgraded. There is no upper limit for this.

$$\sum_{tn \leq t} TCADD_{tl,tn,r,allr} \leq tl1up_{r,allr}, \qquad (3.16)$$

r = Germany, France, Italy, U.K., Belgium/Netherlands, Spain/Portugal;
$allr$ = Germany, France, Italy, U.K., Belgium/Netherlands, Spain/Portugal, Scandinavia, Eastern Europe;
tl = $tl1up$.

$$\sum_{tn \le t} TCADD_{tl,tn,r,allr} \le tl2up_{r,allr}, \tag{3.17}$$

$r =$ Germany, France, Italy, U.K., Belgium/Netherlands, Spain/Portugal
$allr =$ Germany, France, Italy, U.K., Belgium/Netherlands, Spain/Portugal, Scandinavia, Eastern Europe;
$tl = tl2up$.

Power imports from satellite regions are restricted as well:

$$\sum_{s}\sum_{b}\sum_{r} PIMP_{s,b,t,satr,r} \cdot duration_{b,s} \le imup_{t,satr}, \tag{3.18}$$

$t = 1985, \ldots, 2055$;
$satr =$ Scandinavia, Eastern Europe.

3.8 Environmental Taxes

The model permits an analysis of the effects of taxing energy inputs and CO_2 emissions. $ECTAX_{tr}$ measures the total tax revenue from the combined energy/CO_2 tax proposed by the European Commission. The parameters *tecCO2*, *tecMWh* and *tecHYD* stand for the different tax rates and the parameter *co2fac* represents the fuel-specific CO_2-emission:

$$\begin{aligned} ECTAX_{t,r} &= tecCO2_{t,r} \cdot \sum_{j}\sum_{v}\sum_{s}\sum_{bp} co2fac_{j} \cdot \frac{PRODUC_{j,v,s,bp,t,r} \cdot dur_{s,bp}}{eta_{j,v,r}} \\ &+ tecMWh_{t,r} \cdot \frac{8760}{1000000} \cdot \sum_{johydro}\sum_{v}\sum_{s}\sum_{bp} PRODUC_{johydro,v,s,bp,t,r} \cdot dur_{s,bp} \\ &+ tecHYD_{t,r} \cdot \frac{8760}{1000000} \cdot \sum_{hydro}\sum_{v}\sum_{s}\sum_{bp} PRODUC_{hydro,v,s,bp,t,r} \cdot dur_{s,bp} \end{aligned} \tag{3.19}$$

4 NEWAGE

4.1 Overview

The version of NEWAGE[1] used for this study is an extension of an earlier static model designed to investigate the economic implications of alternative CO_2 abatement strategies for the European Union (Böhringer, Ferris, Rutherford 1996). The new features include a hybrid specification of the production side, which provides the option to represent selected sectors by (linear or nonlinear) activity analysis with a choice among discrete technologies, while characterizing other production sectors by means of flexible neoclassical functional forms. The incorporation of activity analysis enhances the credibility of policy-induced technological adjustments, because substitution possibilities are based on real processes rather than abstract production functions ('nebulous' substitution elasticities).

This chapter provides a detailed technical summary of the model and is organized as follows. Section 4.2 introduces the complementarity format as a flexible mathematical approach to formulate general equilibrium problems. Section 4.3 describes how this format accommodates the hybrid specification of production possibilities. Finally, section 4.4 presents the algebraic model formulation.

4.2 General Equilibrium and the Complementarity Format

The standard Arrow-Debreu equilibrium is determined by three classes of inequality conditions: supply minus demand for every commodity is non-negative (market clearance), no sector earns a positive profit (zero-profit), and income restricts expenditure for the consumer (income balance). There are three vectors of decision variables associated with the solution of the equilibrium conditions: non-negative prices, activity levels (production indices) and consumer incomes. In equilibrium, Walras' law implies complementarity between decision variables and inequality

[1] NEWAGE is an acronym for *N*ational, *E*uropean and *W*orld-wide *A*pplied *G*eneral *E*quilibrium modeling system. This system includes a set of inter-related general equilibrium models, which allows for the consistent economic analysis of policy issues at different spatial levels and time treatments (for a description see Böhringer 1996a).

conditions.[2] First, a commodity price exhibits complementary slackness with a market clearance condition, i.e. there are zero prices on commodities, which are in excess supply and positive prices on commodities, for which supplies equal demands. Second, an activity level is complementary to an exhaustion of product constraint, i.e. there are zero excess profits on activities at positive levels, and zero activity levels on unprofitable activities. And third, a consumer income variable is complementary to an income definition equation, i.e. positive income variables are associated with a balanced budget and deviation from income balance implies zero income variables.

Complementary between decision variables and equilibrium conditions motivates the formulation of equilibrium problems in the complementarity format (CP), which incorporates both first-order equilibrium conditions as well as complementarity features explicitly. Algebraically the equilibrium problem in the complementarity format reads as:

Given: $F: R^N \rightarrow R^N,\ \ell, u \in R^N$

Find: $z, w, v \in R^N$

$$\text{s.t. } F(z) - w + v = 0 \qquad \text{(CP)}$$

$$\ell \le z \le u,\ w \ge 0,\ v \ge 0$$

$$w^T(z - \ell) = 0,\quad v^T(u - z) = 0$$

$$\infty \le l \le u \le \infty$$

where:

z	represents a vector of decision variables,
v, w	are vectors of slack variables,
u	is a vector of upper bounds, and
l	denotes a vector of lower bounds

With

$$z = [y, p, Inc]^T \quad \text{and} \quad F(z) = [-A^T p,\ b + Ay - d(p),\ bp - Inc]$$

where:

y is a vector of activity levels,

[2] Note that complementarity is a feature of an economic equilibrium and not an a priori condition.

p	represents a vector of prices,
Inc	denotes a vector of household incomes,
A	describes the economy-wide technology matrix of input-output-coefficients, and
$d(p)$	is a vector of household excess demands.

We obtain the conditions and features of the standard Arrow-Debreu equilibrium by letting $l = 0$ and $u = +\infty$.

The standard Arrow-Debreu framework can be extended to reflect market restrictions such as administered prices and quantities. Within the CP formulation, these price and quantity constraints are related to two additional classes of variables. Endowment rationing instruments accommodate quantity adjustments due to price administrations. Endogenous tax rates indicate the price wedge between the marginal cost and the market price of a commodity caused by quantity administration. Endogenous endowment multipliers exhibit complementary slackness with the quantity constraint, whereas endogenous tax rates are complementary with price constraints.

The advantage of the CP format is that it represents general equilibrium conditions and features in its most general form. A system of inequalities and equations allows for weak inequalities and complementary slackness. This can be used to reflect important properties of a real-world economy such as alternative technical processes for the production of a single commodity or bounds on prices and quantities. Traditional approaches for the formulation of an economic equilibrium as mathematical programming or systems of (nonlinear) equations typically sacrifice the generality of formulation for the ease of computation. Mathematical programming (optimization) accommodates the use of inequalities, but can only be used directly, when the economic problem is integrable (Takayama and Judge 1971). The integrability of the equilibrium problem requires certain regularity conditions, which must be met by the demand functions. These integrability conditions guarantee that the demand functions can be derived from a superior utility maximization problem (-function). The conditions are met, if the Jacobi matrix of the demand functions is symmetric, i.e. the cross-price effects are symmetric.[3] The integrability conditions exclude relevant economic problems with distinct consumer preferences, ad-valorem taxes etc. Sequential joint maximization provides a possible but tedious device to "non-integrability" in the programming approach.[4] Equation methods on the other can deal

[3] The integrability problem is also known as the problem of exact aggregation, which describes the conditions for replacing the demand functions of heterogenous consumers through one single aggregate demand function.

[4] The solution of non-integrable equilibrium problems is enabled through a two-step programming algorithm involving parametric revision and optimization (see e. g. Rutherford 1992).

with non-integrabilities but do not allow for bounds on individual variables nor do they easily accommodate complementarity conditions.[5]

Due to the availability of test and tried solvers, mathematical programming and nonlinear equation methods have been the standard approaches in applied general equilibrium (AGE) modeling despite their restrictions with respect to economic features. Recent commercial developments of algorithms for solving nonlinear complementarity problems provide a powerful alternative to the traditional modeling formats in large-scale AGE modeling and have motivated the formulation of the present NEWAGE model in the CP format.[6]

4.3 Integration of Activity Analysis into AGE Models

In the traditional design, AGE models employ neoclassical nonlinear ("smooth") production functions to represent substitution possibilities across different inputs and transformation possibilities across different outputs. On the input (output) side, these functions give rise to isoquants (transformation curves), which indicate continuous substitutability between inputs (outputs) for a specific level of output (input). Elasticities of substitution (transformation) determine relative changes in quantities due to relative changes in prices. Typically, there are no choices between alternative (nonlinear) production functions capable of producing the same output - a single continuous function represents the production space. The virtue of this approach is a very compact description of technological options at the sectoral or economy-wide level. Disadvantages include the lack of technological foundation and the crucial role of "nebulous" elasticities.

Many system analysts stress the importance of a detailed representation of discrete alternative technological options, which might better assure credibility of policy simulations. This leads to an activity analysis framework (Koopmans 1951), where technologies may be operated or not depending on profitability. It is important to be aware that the activity analysis framework applies to both linear as well as nonlinear technologies. In the activity analysis framework production of a good is typically described by means of a linear technology matrix with a finite number of Leontief production activities, which are represented by vectors of fixed input-output coefficients. In this case, the convex combination of adjacent activities yields an isoquant which is not smooth: the rate of substitution stays constant in a certain range, but then takes an abrupt jump. Hence, linear activity analysis can lead to extreme, rather unrealistic specialization (so called "penny-switching"), when expansion, introduction or phase-out of discrete technologies are not restricted by additional constraints. In economy-wide modeling, the use of activity analysis for the

---

[5] For a general discussion of alternative methods for the formulation of an economic equilibrium see Böhringer (1996b).

[6] Note that the CP format includes the mathematical programming format and the equation based format as special cases (see Rutherford 1994).

specification of consistent technological processes is limited by the availability of concrete technological (engineering) data and dimensionality.

Given the specific advantages and disadvantages, the mixture of both apporaches in a single framework appears to be an attractive modeling strategy: To enhance credibility, those production sectors, whose technological options are of major importance for the policy issue, can be described through linear activity analysis. To restrict data requirements as well as the dimensionality of a complex-world model, the technological options of the remaining production sectors can be described by means of single continuous neoclassical production functions. Mathematical programming allows for the representation of both, discrete alternative technological options and continuous substitution possibilities. However, as indicated in section 4.2, the simple programming approach requires integrability of the equilibrium problem and thus excludes important economic features. On the other hand, equation based approaches, which accommodate non-integrabilites, are not suited for activity analysis. This explains why models based on nonlinear equation methods do not incorporate activity analysis.

General equilibrium conditions for production as stated in the CP format facilitate both, the use of a single equation (smooth production function) as well as the use of inequalities to capture technological alternatives. In the latter case we obtain the typical activity analysis with complementary slackness between the level of activity and the zero-profit condition. For a single smooth production function the activity analysis "degenerates" to a special case where the activity level remains positive throughout the price space.

In AGE modeling activity analysis is hardly incorporated. Equation based approaches, which accommodate the description of non-integrable general equilibrium problems, are not suited for activity analysis, since this involves the specification of inequalities. Mathematical programming, which allows for activity analysis, is too restrictive for many economy-wide problems due to integrability conditions. As compared to other approaches, the CP format states equilibrium conditions in its most general form and provides a straightforward way to incorporate activity analysis in non-integrable equilibrium problems.[7]

Suppose the activity analysis representation of sectoral production activities involves a set of discrete Leontief technologies ($t = 1, ..., s$). Each technology t of sector j is represented through a fixed unit „netput"-vector a_{tj}. The i-th element of this vector a_{itj} represents output minus input of commodity i per unit operation of technology t in sector j.

Accounting for the specific factor $\bar{Q}_{tj}$ (typically fixed capital) in activity analysis representation of a sector yields the following equilibrium conditions and complementarity features:

[7] The following exposition is based on Böhringer (1996c).

$$\left(-\sum_i a_{itj}\; p_i + p_{\bar{Q}tj}\right) \geq 0 \quad \wedge \quad \left(-\sum_i a_{itj}\; p_i + p_{\bar{Q}tj}\right) y_{tj} = 0 \tag{4.1}$$

$$\left(-a_{\bar{Q}tj}\; y_{tj} + Q_{tj}\right) \geq 0 \quad \wedge \quad \left(-a_{\bar{Q}tj}\; y_{tj} + \bar{Q}_{tj}\right) p_{\bar{Q}tj} = 0 \tag{4.2}$$

Expression (4.1) states the zero-profit condition at the single technology level (including pure profits) and complementary slackness between production and profitability. Expression (4.2) corresponds to the market clearance condition for the specific factor of a technology $\bar{Q}_{tj}$ and relates a binding capacity constraint on active technologies to a positive capacity rent $p_{\bar{Q}tj}$.

Figure 4.1. illustrates a step-wise supply function, which emerges from the specific factors $\bar{Q}_{tj}$ setting an upper bound on production per technology (capacity constraint).

Figure 4.1. Step-Wise Supply Curve in Bottom-Up Approach

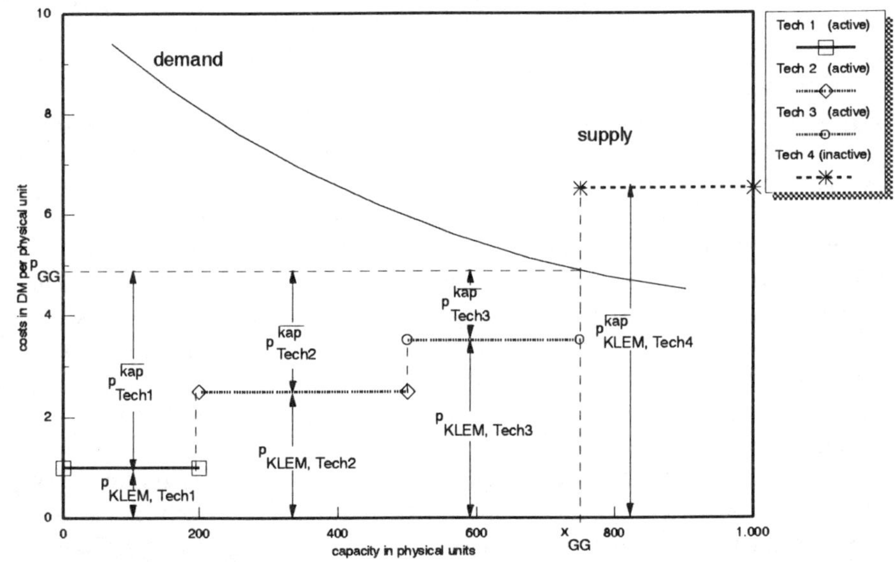

Ownership of the fixed factor determines the allocation of pure profits (rents) from the sector. The rent on specific capital $p_{\bar{Q}tj}$ can be visualized as the difference between the market price p_{GG} and the aggregate input costs $p_{Agg,t}$ of a technology (net of capacity rents). Those technologies, which are active in the benchmark break even (i.e. make zero profits gross of capacity rents) and those technologies, which are inactive, would run an economic loss at prevailing prices. Policy interference (e.g. the imposition of taxes) changes the relative price system. Due to the change in relative prices inactive technologies might turn active and formerly active technologies may turn inactive.

4.4 Algebraic Summary[8]

This section provides an algebraic summary of equilibrium conditions for the single-period multi-sector, multi-region submodel of NEWAGE in which:

(1) Output and factor prices are fully flexible and markets are perfectly competitive.

(2) Labor is inter-sectorally mobile within a region, but cannot move between regions. Capital rents accrue to both sector-specific capital and malleable capital. The latter is freely mobile across sectors and countries.

(3) In international trade, goods are differentiated by region of origin (Armington 1969). Regions are linked together through endogenous bilateral trade flows calibrated to base year export and import values. A set of export demand and import supply functions determines the trade between ROW (*R*est *O*f the *W*orld) regions, which are not explicitly represented and those countries whose production and consumption patterns are described in detail.

(4) Government demand within each region, as well as the balance of payment surplus, are fixed at exogenous levels. Investment demand (savings) is determined through a constant marginal propensity to save by private households.

(5) There is one representative consumer for each region.

The model equations correspond to the three classes of conditions associated with the Arrow-Debreu general equilibrium: first, exhaustion of product (zero-profit) conditions for constant-returns to scale producers; second, market clearance for all goods and factors (including emission permits for the case of emission abatement) and trade balance with ROW; and third, income balance for representative agents in each of the explicitly modelled regions. An equilibrium allocation determines market production, prices and incomes. In the algebraic exposition below, the notation Π_{jr}^{X} is used to denote the profit function of sector j in region r; where X is the name assigned to this activity. Formally, all production sectors exhibit constant returns to scale, hence, differentiating Π_{jr}^{X} with respect to input and output prices provides compensated demand and supply coefficients, which appear subsequently in market-clearance conditions.

4.4.1 Exhaustion of Product Conditions

Within each region nested, separable, constant elasticity of substitution (CES) cost functions with three levels are employed to specify the substitution possibilities in

[8] Sections 4.4.1 - 4.4.3 are based on Böhringer, Ferris and Rutherford (1996). The model is formulated as a nonlinear system of inequalities using GAMS/MPSGE (Rutherford 1994) and solved using PATH (Dirkse and Ferris 1995).

domestic production between capital, labor, energy and material inputs (KLEM). The material input of good i in sector j corresponds to an Armington aggregate of non-energy inputs from domestic production and imported varieties, which trade off with an constant elasticity of substitution (in equation (4.3) below the index EG denotes the set of energy goods). At the top level these material inputs are employed in fixed proportions with an aggregate of energy, capital (sectorally fixed and regionally mobile) and labor. A constant elasticity describes the substitution possibilities between the energy aggregate and the aggregate of labor and capital at the second level. Finally, at the third level labor and capital trade off with a constant elasticity of substitution.[9] The resulting zero profit condition for the production of good j in region r is:

$$\Pi^{Y}_{jr} = P_{jr} - \sum_{i \notin EG} a_{ijr} \left(\theta^{M}_{ijr} \, P^{M\;1-\sigma_{DM}}_{ir} + (1-\theta^{M}_{ijr}) \, P^{1-\sigma_{DM}}_{ir} \right)^{\frac{1}{1-\sigma_{DM}}}$$

$$- a^{KLE}_{jr} \left[\alpha^{E}_{jr} \, P^{E\;1-\sigma_{KLE}}_{jr} + (1-\alpha^{E}_{jr}) \left(w_r^{\beta^{L}_{jr}} \, r_{jr}^{\beta^{S}_{jr}} \, R^{\beta^{K}_{jr}} \right)^{1-\sigma_{KLE}} \right]^{\frac{1}{1-\sigma_{KLE}}} = 0 \qquad (4.3)$$

where:

P_{jr}	is the output price of good j produced in region r,
a_{ijr}	is for the benchmark value share of non-energy input i in sector j of region r,
Π_{jr}	is the import value share for sector i inputs to sector j in region r,
P^{M}_{ir}	is the import price aggregate for good i imported to region r,
σ_{DM}	is the elasticity of substitution between domestic and imported inputs or demands,
a^{KLE}_{jr}	corresponds to the aggregate value share of capital, labor and energy inputs (KLE aggregate) in sector j of region r,
α_{jr}	denotes the energy input value share of the KLE aggregate in sector j of region r,
P^{E}_{jr}	stands for the composite price for aggregate energy inputs into sector j in region r,
σ_{KLE}	represents the elasticity of substitution between the energy aggregate and the aggregate of capital and labor,
w_r	is the economy-wide wage rate in region r,

[9] To assure consistency with the nesting structure in LEAN the zero-profit condition for sectoral production (4.3) of the base model is changed as follows: At the second level labor trades off with an aggregate of capital and energy subject to a constant elasticity of substitution. At the third level, a constant elasticity describes the substitution possibilities between capital and energy.

r_{jr} is the rate of return for sector-specific capital inputs in sector j of region r,

R is the uniform rate of return for (inter-regionally and inter-sectorally) mobile capital,

β_{jr}^{k} denotes the value shares for labor ($k = L$), sector-specific capital ($k = S$) and inter-regionally mobile capital ($k = K$) in sector j of region r,

Y_{jr} is the associated dual variable which indicates the activity level of producing good j in region r.

Intermediate demands for fuels are split into energy throughput ('non-energetic') and energy use ('energetic') components. The first of these demands enters through the material aggregate, while the second forms the energy input aggregate. Energetic energy demands are specified by means of a two-level CES function. At the bottom level an Armington aggregation function provides a constant elasticity of substitution between domestic and imported varieties of each energy good. The resulting Armington energy good is linked to a specific carbon emission coefficient, which determines the carbon tax payment for this energy carrier. At the top level different Armington energy goods (after accounting for carbon tax payments) trade off with a constant elasticity of substitution. The zero profit condition for energy supply to sector j in region r is given by:

$$\Pi_{jr}^{E} = P_{jr}^{E} - \Big[\sum_{i \in EG} \theta_{ijr}^{E} \Big(\big(\theta_{ijr}^{M} \; P_{ir}^{M\;1-\sigma_{DM}} + (1-\theta_{ijr}^{M}) \; P_{ir}^{1-\sigma_{DM}} \big)^{\frac{1}{1-\sigma_{DM}}} + \varepsilon_{ijr} \; PCO \Big)^{1-\sigma_{E}} \Big]^{\frac{1}{1-\sigma_{E}}} = 0 \qquad (4.4)$$

where:

P_{jr}^{E} is the composite price for aggregate energy inputs into sector j in region r,

θ_{ijr}^{E} is the benchmark value share of energy good i in aggregate energy demand by sector j of region r,

ε_{ijr} is the carbon emission coefficient for energy input i in sector j of region r,

σ_E is the elasticity of substitution between energy inputs[10],

---

[10] Equation (4-4) as well as equation (4-9) are simplifications. In the current model formulation, one additional level of nesting for energy inputs is introduced, so that electric and non-electric energy first trade off with a constant elasticity of substitution, and subsequently fossil energy inputs (oil, gas and coal) enter in a lower CES nest. Substitution possibilities in electricity generation between different energy inputs

PCO is the market price of carbon emission rights,

E_{jr} is the associated activity level representing aggregate energy demand for producing good j in region r.

Import supply for region r and good j is composed of a CES-aggregate of imports from all other regions (including ROW). The exhaustion of product condition for these import activities is given by:

$$\Pi_{jr}^{M} = P_{jr}^{M} - \left(\sum_{r' \neq r} \theta_{j'rr}^{MM} \; P_{j'r}^{1-\sigma_{MM}} + \theta_{jROWr}^{MM} \; P_{jROW}^{M1-\sigma_{MM}} \right)^{\frac{1}{1-\sigma_{MM}}} = 0 \qquad (4.5)$$

where:

$\theta_{j'rr}^{MM}$ is the benchmark value share of region r' exports in aggregate imports of good i into region r,

θ_{jROWr}^{MM} is the benchmark value share of ROW exports in aggregate imports of good i into region r,

σ_{MM} is the elasticity of substitution between imports from different foreign countries,

P_{ROW} is the price for good j produced in ROW,

M_{jr} is the associated activity level of this constraint, meaning the level of demand in region r for the aggregate import variety of commodity j.

In each region composite investment is a Leontief aggregation of Armington inputs, which are composed of domestic and imported commodities:

$$\Pi_{r}^{I} = P_{r}^{I} - \sum_{i} a_{ir}^{I} \left(\theta_{iIr}^{M} \; P_{ir}^{M1-\sigma_{DM}} + (1-\theta_{iIr}^{M}) \; P_{ir}^{1-\sigma_{DM}} \right)^{\frac{1}{1-\sigma_{DM}}} = 0 \qquad (4.6)$$

where:

P_{r}^{I} stands for the composite price for investment demand in region r,

a_{ir}^{I} is the benchmark value share of input i in investment of region r,

θ_{iIr}^{M} denotes the import value share for sector i inputs into investment formation of region r,

(equation 4.4) and between energy and other inputs (equation 4.3) are captured by linear activity analysis (see section 4.4.4)

I_r is the associated activity level, representing investment supply in region r.

Public goods and services are produced with a Cobb-Douglas aggregation of commodity inputs, in which an Armington aggregate of domestic and imported commodities is embedded:

$$\Pi_r^G = P_r^G - \prod_i \left(\theta_{iGr}^M \, P_{ir}^{M^{1-\sigma_{DM}}} + (1-\theta_{iGr}^M) \, P_{ir}^{1-\sigma_{DM}} \right)^{\frac{\alpha_{ir}^G}{1-\sigma_{DM}}} = 0 \tag{4.7}$$

where:

P_r^G is the composite price for government demand in region r,

θ_{iGr}^M denotes the import value share for sector i inputs into public good production of region r,

α_{ir}^G represents the value share of commodity i in public output of region r,

G_r is the level of public output in region r.

Household demand is given as a CES composite, which combines consumption of an energy aggregate and a non-energy consumption bundle. Substitution patterns within the non-energy consumption bundle are Cobb-Douglas functions with an Armington aggregation of imports and domestic commodities. Non-energy Armington goods are an aggregate of domestic and imported varieties, which trade off with a constant elasticity of substitution. Exhaustion of production for household demand is given by:

$$\Pi_r^C = P_r^C - \Big[\theta_{Cr}^E \, P_{Cr}^{E^{\,1-\sigma_{EC}}} + (1-\theta_{Cr}^E) \Big(\prod_{i \notin EG} \big(\theta_{iCr}^M \, P_{ir}^{M^{1-\sigma_{DM}}} + (1-\theta_{iCr}^M) \, P_{ir}^{1-\sigma_{DM}} \big)^{\frac{\alpha_{ir}^C}{1-\sigma_{DM}}} \Big)^{1-\sigma_{EC}} \Big]^{\frac{1}{1-\sigma_{EC}}} = 0 \tag{4.8}$$

where:

P_r^C stands for the composite price for aggregate household demand in region r,

θ_{Cr}^E represents the benchmark value share of the energy aggregate in household demand of region r,

P_{Cr}^E is the composite price for aggregate energy inputs into household demand in region r,

σ_{EC} is the elasticity of substitution between the energy aggregate and the non-energy consumption bundle in household demand,

θ^M_{iCr} denotes the import value share for sector i inputs into household demand of region r,

I_r corresponds to the value share of non-energy commodity i in household demand for non-energy consumption bundle of region r

C_r is the associated activity level representing aggregate household consumption in region r.

Energy consumption by households is characterized in the same way as energy demand of producing sectors, with zero profit condition being written as:

$$\Pi^E_{Cr} = P^E_{Cr} - [\sum_{i \in EG} \theta^E_{iCr} \; ((\theta^M_{iCr} \; P^{M \; 1-\sigma_M}_{ir} + (1-\theta^M_{iCr}) \; P_{ir}^{1-\sigma_M})^{1-\sigma_M} + \varepsilon_{iCr} \; PCO \;)^{1-\sigma_E} \;]^{\frac{1}{1-\sigma_E}} = 0 \tag{4.9}$$

where:

θ^E_{iCr} represents the benchmark value share of energy good i in aggregate energy demand by the household in region r,

ε_{iCr} corresponds to the carbon emission coefficient for energy Input i into household demand of region r[11]

E_{cr} is the associated activity level of household energy demand in region r.

4.4.2 Market Clearance

In this exposition Shepard's Lemma is exploited to provide a compact representation of compensated demand and supply functions. Primary factors of production are labor, sector-specific capital for each region and inter-regionally mobile capital. The market clearance conditions for labor are written:

$$\bar{L}_r = \sum_j Y_{jr} \frac{\partial \, \Pi^Y_{jr}}{\partial \, w_r} \tag{4.10}$$

where:

$\bar{L}_r$ is the aggregate labor endowment for region r.

---

[11] As with industries households have to buy emission rights (i.e. pay taxes) in order to use carbon emitting inputs within their energy consumption bundle.

The supply-demand balance for inter-regionally mobile capital is written:

$$\sum_r \overline{K}_r = \sum_{j,r} Y_{jr} \frac{\partial \Pi^Y_{jr}}{\partial R} \tag{4.11}$$

where:

$\overline{K}_r$ denotes the aggregate endowment of inter-regionally mobile capital for region r.

As to sector-specific capital, market clearance corresponds to:

$$\overline{K}^S_{jr} = Y_{jr} \frac{\partial \Pi^Y_{jr}}{\partial r_{jr}} \tag{4.12}$$

where:

$\overline{K}^S_{jr}$ is the sector-specific capital for sector j in region r.

Goods produced in each region enter intermediate demand, consumer demand, government and investment demand as well as import demand from other regions including ROW. The market clearance condition for each produced commodity is written:

$$Y_{ir} = \sum_j Y_{jr} \frac{\partial \Pi^Y_{jr}}{\partial P_{ir}} + C_r \frac{\partial \Pi^C_r}{\partial P_{ir}} + I_r \frac{\partial \Pi^I_r}{\partial P_{ir}} + G_r \frac{\partial \Pi^G_r}{\partial P_{ir}} + \sum_{r'} M_{i'r} \frac{\partial \Pi^M_{i'r}}{\partial P_{ir}} + M_{irROW} \tag{4.13}$$

where:

M_{irROW} is the aggregate export demand of ROW for good i from region r (see the foreign closure rule in equation (4.15) below).

The market for imports is analogous to the market for regional outputs. The supply-demand balance for imported goods is written:

$$M_{ir} = \sum_j Y_{jr} \frac{\partial \Pi^Y_{jr}}{\partial P^M_{ir}} + C_r \frac{\partial \Pi^C_r}{\partial P^M_{ir}} + I_r \frac{\partial \Pi^I_r}{\partial P^M_{ir}} + G_r \frac{\partial \Pi^G_r}{\partial P^M_{ir}} \tag{4.14}$$

As to the trade balance of regions with respect to the ROW, a simple foreign closure rule is employed. The Armington assumption of product heterogeneity is used along with a ROW export-demand function of constant price elasticity, price-taking behavior of countries with respect to world import (ROW) prices (perfectly elastic foreign import-supply function), and an imposed balance of payment constraint to assure trade balance between single regions and ROW:

$$\sum_{i,r} \overline{P}^x_i \; M_{irROW} = \sum_{i,r} \overline{P}^M_i \; M_{iROWr} + \sum_r \overline{B}_r \tag{4.15}$$

where:

- B_r is the net trade (balance of payment) surplus for region r,
- P^M_i are the ROW prices of imports M_{iROWr} from ROW to regions r,
- P^X_i are the ROW prices of exports M_{irROW} of region r to ROW.[12].

The supply-demand balance for carbon emission rights is written:

$$\sum_r CRTS_r = \sum_{j,r} Y_{jr} \frac{\partial \Pi^Y_{jr}}{\partial PCO} \tag{4.16}$$

where:

- $CRTS_r$ indicates the carbon emission rights endowment of region r.

4.4.3 Income and Aggregate Demand

Total income of the representative household in each region is employed in fixed fractions on household demand and savings (constant marginal propensity to save). Given aggregate price indices for consumption and investment there is:

[12] The associated (dual) variable with this constraint indicates the real exchange rate relative to ROW (e.g., 1985 ECU/1985 US$).

$$C_r = (1 - mps_r)(w_r \; \overline{L}_r + R \; \overline{K}_r + \sum_j r_{jr} \; K^S_{jr} + PCO \; CRTS_r - P^G_r \; \overline{G}_r - \overline{B}_r) / P^C_r \tag{4.17}$$

$$I_r = mps_r(w_r \; \overline{L}_r + R \; \overline{K}_r + \sum_j r_{jr} \; K^S_{jr} + PCO \; CRTS_r - P^G_r \; \overline{G}_r - \overline{B}_r) / P^C_r \tag{4.18}$$

where:

mps_r is the constant marginal propensity to save in region r.

4.4.4 Activity Analysis Representation of the Electricity Sector

Policy simulations on the economic effects of alternative CO_2 abatement strategies for the EU have motivated an activity analysis representation of the electricity sector in the current model. Among all industries the electricity sector exhibits the highest potential for CO_2 mitigation through fuel-switching and energy efficiency improvements. Hence, a technological foundation of carbon substitution possibilities in this sector is desirable to strengthen the credibility of policy-induced technological choice. As to the activity analysis representation of the electricity generation the specification of a few key technologies is sufficient to give a realistic picture of available technological options. *Table 4.1.* summarizes the set of discrete technologies considered in the model (see section 5.1.1 for the abbreviations of technologies).

Table 4.1. Summary of Discrete Technologies for Electricity Generation

Technology	load type
nuclear (NUC)	base
soft coal (BC)	base
hard coal (HC)	middle
hydro (HYDRO_L)	middle
gas (GASTUR)	peak
oil (OIL)	peak
gas_chp1 (GAS_CC)*	base
gas_chp2 (GAS_CC)	middle
hard coal- NT** (HC_CC)	middle
wind	middle
solar	middle

*chp: combined heat and power
**NT: new blast furnace technology

Within one load type technologies generate perfectly substitutable electricity. Electricity of different load types are merged together through a CES technology producing a composite electricity good, which is demanded from the rest of economy:[13]

$$y_{ELE,r} = \left(\sum_{l=ba,mi,pe} \nu_{lr} \, \psi_{l,ELE,r}^{\frac{\eta_{ELE,r}-1}{\eta_{ELE,r}}} \right)^{\frac{\eta_{ELE,r}}{1-\eta_{ELE,r}}} \tag{4.19}$$

where:

$y_{ELE,r}$	is the activity level of electricity production in region r,
l	is the index to denote the load type (*ba* - base, *mi* - middle, *pe* - peak),
ν_{lr}	denotes the load-specific distribution parameter in region r,
$y_{l,ELE,r}$	is the activity level of electricity of load type l in region r,
$\eta_{ELE,r}$	indicates the elasticity of substitution between different load types in region r.

[13] The aggregation is necessary because input-output tables and trade statistics report electricity flows only in terms of a homogeneous good.

Figure 4.2. illustrates the generic structure of electricity production in a network representation where boxes symbolize technologies, vertical lines indicate sinks and sources of commodities, and horizontal lines indicate commodity in- and outflows. Depending on the input structure and the relative price system, technologies are profitable or not. The maximum activity level of a technology is restricted by capacity bounds, which reflect expert estimates on the scope of technology expansion during the time horizon.

Figure 4.2. Network Representation of Electricity Production

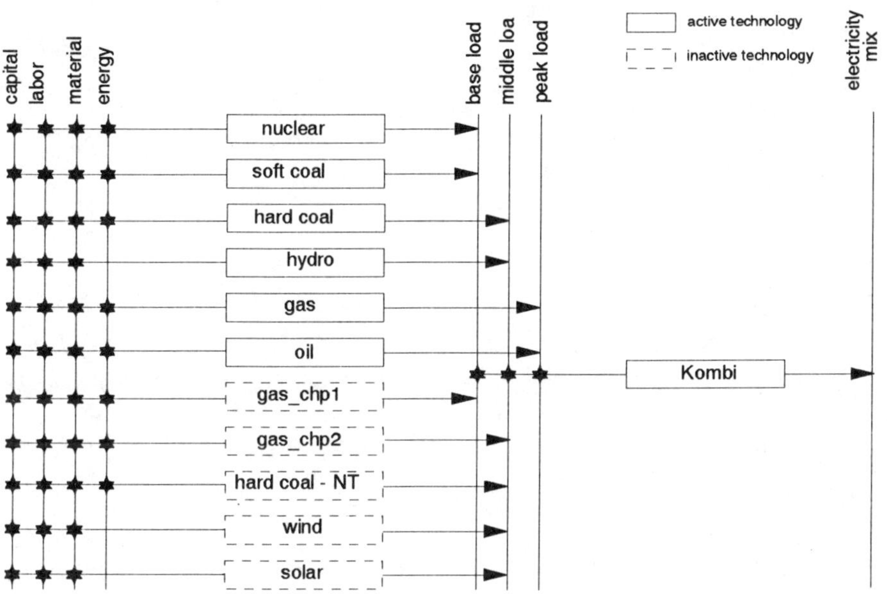

5 Parametric Framework and Policy Scenarios

5.1 Parametric Framework

5.1.1 Regional, Sectoral and Technological Disaggregation

The models used in this study employ a common disaggregation of regions, economic sectors and electricity generation technologies.

The regions[1] distinguished are the following:

GE	Germany
FR	France
UK	United Kingdom
IT	Italy
BN	Belgium/Netherlands
IB	Spain/Portugal

The production sectors (identified with homogeneous goods) include:

AGR	Agriculture
HC	Hard coal
BC	Lignite/brown coal
OIL	Crude oil and oil products
GAS	Natural gas and manufactured gas
EL	Electricity and steam
BAS	Basic materials
EQU	Equipment goods
CON	Consumption goods
CST	Construction
TRA	Transportation
PRS	Private services
PUB	Public services

The definition of sectors in terms of the EUROSTAT three digit sector labels is shown in *Table 5.1*.

---

[1] Throughout the study, regions are listed in the order of their GDP.

Table 5.1. Aggregation of Production Sectors from R59 EUROSTAT Input-Output Table (three digits indicate the R59 labeling index)

AGR	010 Agricultural & forestry & fishery products, 095 Water
HC	031 Coal, 050 Products of coking
BC	033 Lignite
OIL	071 Crude oil, 073 Refined petroleum products
GAS	075 Natural gas, 098 Manufactured gases
EL	097 Electric power, 099 Steam
BAS	110 Nuclear fuels, 135 Iron ore ECSC iron and steel products, 136 Non ECSC iron and steel products, 137 Non-ferrous metals, 151 Cement lime & plaster, 153 Glass, 155 Earthen ware and ceramic products, 157 Other mineral and derived products, 170 Chemical products
EQU	190 Metal products, 210 Agricultural & industrial machinery, 230 Office machines, 250 Electrical goods, 270 Motor vehicles & engines, 290 Other transport equipment
CON	310 Meat & meat products, 330 Milk & dairy products, 350 Other food, products, 370 Beverages, 390 Tobacco products, 410 Textiles & clothing, 430 Leather & footwear, 450 Timber & wooden furniture, 471 Pulp & paper & board, 473 Paper goods and products of printing, 490 Rubber & plastic products, 510 Other manufacturing products
CST	530 Building & civil engineering works, 550 Recovery & repair services
TRA	611 Railway transport services, 613 Road transport, 617 Inland waterway services, 631 Maritime & coastal transport services, 633 Air transport services, 650 Auxiliary transport services
PRS	570 Wholesale & retail trade, 590 Lodging & catering services, 670 Communications, 690 Credit & insurance, 710 Business services provided to enterprises, 730 Renting of immovable goods, 750 Market services of education & research, 770 Market services of health, 790 Other Market services,
PUB	810 General public services, 850 Non-market services of education & research, 890 Non-market services of health, 930 Other Non-market services

The following electricity generation technologies are distinguished:

HC	Hard coal (conventional technology)
HC-CC	Hard coal combined-cycle technology
HCGAS	Bivalent generation with hard coal and gas
BC	Brown coal (conventional technology)
OIL	Oil (conventional technology)
GAS-CC	Gas combined-cycle technology
GASTUR	Gas turbine generation
NUC	Nuclear power
HYDRO-L	Run of river
HYDRO-P	Storage and pumped storage

5.1.2 Benchmark Data and Elasticities for LEAN and NEWAGE

As is customary in applied general equilibrium analysis, the models LEAN and NEWAGE are based on economic transactions in a particular benchmark year (1985 in this case). Benchmark data determine parameters of the functional forms from a given set of benchmark quantities, prices and elasticities (so-called calibration). The basic data set of the general equilibrium models consist of input-output tables, trade matrices and data on investment and capital stocks for the benchmark year 1985.

The input-output data base covers the outputs and intermediate inputs of the 13 sectors mentioned in subsection 5.1.1, the primary inputs (labor, profits, depreciation and production taxes), imports, and the final demand categories (consumption, investment, government expenditures and exports). Prices in the benchmark year are normalized to unity; thus values and quantities in the benchmark year coincide. The general quantity unit in the model is 'billion ECU at 1985 prices'.

The raw input-output data are from EUROSTAT (1992). The choice of 1985 as the benchmark year stems from the fact that a consistent European input-output data base is being prepared only with a considerable time delay. In fact, a slight deviation should be noted, as data for Portugal are available only for 1980, and the raw data for the Netherlands refer to 1987. They have been scaled to 1985 using data from the National Accounts.

For the implementation of LEAN, sectoral capital stocks and a capital composition matrix were required (the latter showing the composition of each sector's investment in terms of sectoral origin). This data base has been constructed on the basis of raw data from Gerstenberger et al. (1989) for Germany and EUROSTAT (1991) for the other regions.

Important model parameters are the elasticities of substitution, and the reaction elasticities of the savings ratio, the wage rates and the exchange rate.
The elasticities of substitution follow estimates found in the relevant literature[2]. For foreign trade, elasticities are differentiated by type of good. The

[2] For an overview see Burniaux et al. (1992).

differentiation has been guided by considerations of transport costs, the availability of transmission networks (if applicable) and institutional restrictions. The existence of domestic sources of the good is also considered essential for substitution elasticities. In Germany, an example of rare domestic sources is natural gas. Clearly enough, a world market for natural gas fails to exist due to high costs of shipment for liquified natural gas. Thus, the natural gas market is primarily a continental one, with a low substitution elasticity. For brown coal, world trade does not pay and, hence, does not exist. For obvious reasons there is also no world trade in electricity. In total, it is evident that non-market services are poor international substitutes.

In all cases in which international trade, be it on a world or European scale, is non-existent or strongly restricted, we have fixed the corresponding substitution elasticity at 0.1. Otherwise, the default value for exports of the EU versus those of the rest of the world is set at 0.8 and the one for exports of one EU region versus those of another is set at 4.0. For imports versus domestic production the default value is fixed at 2.0.

For production we have assumed uniform elasticities across all sectors.[3] The values for energy-capital versus labor and for energy versus capital are those employed by Burniaux et al. (1992). Within the energy-capital aggregate we follow the idea that substitution elasticities become smaller as aggregates become broader. Probably most remarkable is the low value for energy-capital substitutability. However, the figure employed is in line with other studies, e.g., Manne and Richels (1992). In fact, a range of 0.3 to 0.4 may be regarded as a 'consensus outcome' of the long-lasting substitutability-complementarity controversy.

The elasticities in the reaction functions for the savings ratio, the exchange rate and the wage rate have been chosen quite informally, with a view to long-term stability. For instance, if there is capital-augmenting technical progress a strictly positive elasticity of the savings ratio with respect to the interest rate turns out to be necessary in order to keep the interest rate in a reasonable range. Similarly, a reasonable long-term behavior of the current account balance requires sufficient exchange rate flexibility. Finally, it is plausible to assume that the elasticity of the wage rate with respect to employment is above unity (otherwise, more labor could be mobilized at no incremental cost).

The elasticity assumptions are collected in *Table 5.2*.

---

[3] This is an obvious candidate for future improvement.

Table 5.2. Elasticity Assumptions

Substitution Elasticities	
Exports ROW/Exports EU	0.1-0.8
Exports EU Region/Exports EU Region	0.1-4.0
Imports/Domestic Production	0.1-2.0
Energy-Capital/Labor	0.6
Energy/Capital	0.5
Electricity/Fuel	0.8-1.2
Fuel/Fuel	0.8
Reaction Elasticities	
Savings Ratio	0.8
Wage Rate	1.5
Exchange Rate	0.3

In NEWAGE traded goods are distinguished by place of production (EU member countries and rest of the world). The share parameters in the function specification for a good i supplied by region r and demanded in a country are calculated using a trade matrix with bilateral trade flows. The trade matrices are based on the CHELEM harmonized accounts on trade and world economy (WEFA 1995). Reconciliation of EUROSTAT input-output data and trade data is an important step in constructing the base year equilibrium. For this purpose, we employ a nonlinear least squares procedure to calibrate bilateral trade flows provided by CHELEM to the intra-EU and extra-EU trade totals provided in the Eurostat input-output tables.[4]

5.1.3 Characterization of Technologies in EIREM and NEWAGE

The stock of power plants in the model regions is differentiated by technology (see subsection 5.1.1) and vintage. It covers the power stations in the public supply system coming into operation after 1950 and still available in 1995.

The set of supply technologies captured in the model represents those technologies that may be expected to be in operation on a large scale within the

4 CHELEM does not cover all trade flows, and for the missing goods a target bilateral trade matrix representing the average over all traded goods is used.

time horizon of the model.[5] These technologies are characterized by the following parameters:

- heat rate (percent)
- investment costs (DM/kW)
- specific labor costs (DM per MW and year)
- specific maintenance and repair costs (DM per MW and year)
- other variable costs (Pf per kWh)
- technical lifetime (years)
- average seasonal availability (percent of installed capacity)
- period of depreciation (years)

For these parameters both historical and expected future values have been estimated.

The expected future development of *heat rates* (power plant efficiencies) up to 2020 is shown in *Table 5.3.* The data have been constructed on the basis of information provided by utilities and suppliers of power stations. Heat rates are expected to improve continuously. Conventional hard-coal-fired plants will increase their efficiency advantage over lignite-fired plants, and combined-cycle hard-coal-fired plants will develop an advantage of 1.5 to 2 percentage points over conventional ones. Gas-fired combined-cycle stations will have the highest heat rate. It is expected to approach 63 percent by 2020.

Table 5.3. Heat Rates (Percent)

	1995	2000	2005	2010	2015	2020
HC	42.5	45.5	46.5	47.5	48.5	49.5
HC-CC	45.8	45.8	48.0	49.0	50.0	51.0
HCGAS	45.4	45.5	48.0	49.0	50.0	51.0
BC	41.5	43.0	44.5	45.5	46.5	47.5
OIL	44.0	44.5	45.0	47.0	47.2	47.4
GAS-CC	57.5	60.0	61.5	62.0	62.5	63.0
GASTUR	38.0	38.5	39.0	39.5	40.0	40.5
NUC	34.0	36.0	36.0	36.0	36.0	36.0

The symbols denoting the technologies are defined in subsection 5.1.1.

Investment costs show considerable dispersion across regions. This stems from differences in technological specifications and in the regulatory environment (licensing procedures, safety standards). It is assumed that these differences will

[5] Lignite gasification technology was not taken into consideration due to comparatively high investment costs and negligible improvement in efficiency.

be sustained in a scenario of closed national markets, whereas in the open market case competitive pressure will lead to partial convergence of costs towards 'best practice'. More specifically, in the Internal Market scenarios the excess of investment costs over best practice is assumed to be reduced 70 percent by 2005 and 90 percent by 2010. Best practice investment costs are assumed to be constant over time. This means that future improvements in power plant efficiency have no influence on investment costs.

Investment costs in Germany are shown in *Table 5.4* (closed market case). They have been calculated on the basis of industry information.

Table 5.4. Investment Costs in Germany (DM/kW)

	1995	2000	2005	2010	2015	2020
HC	2250	2250	2225	2201	2201	2201
HC-CC	2800	2800	2769	2738	2738	2738
HCGAS	2200	2200	2176	2152	2152	2152
BC	2500	2500	2490	2480	2480	2480
OIL	1850	1850	1819	1787	1787	1787
GAS-CC	1150	1150	1129	1109	1109	1109
GASTUR	650	650	638	627	627	627
NUC	4300	4300	4090	3879	3879	3879

Investment costs are expressed at 1995 price level.

For the most important technologies, investment costs in the other model regions are displayed in *Table 5.5*, relative to German investment costs (closed market case).

Table 5.5. Investment Costs Relative to German Investment Costs

HC	1995	2000	2005	2010	2015	2020
FR	0.93	0.93	0.94	0.94	0.94	0.94
UK	0.89	0.89	0.90	0.91	0.91	0.91
IT	0.92	0.92	0.93	0.94	0.94	0.94
BN	0.93	0.93	0.94	0.94	0.94	0.94
IB	0.98	0.98	0.98	0.98	0.98	0.98

OIL	1995	2000	2005	2010	2015	2020
FR	0.85	0.85	0.86	0.88	0.88	0.88
UK	0.87	0.87	0.88	0.89	0.89	0.89
IT	0.92	0.92	0.93	0.94	0.94	0.94
BN	0.83	0.83	0.84	0.86	0.86	0.86
IB	0.93	0.93	0.94	0.95	0.95	0.95

Table 5.5. (continued)

GAS-CC	1995	2000	2005	2010	2015	2020
FR	0.85	0.85	0.86	0.88	0.88	0.88
UK	0.85	0.82	0.84	0.85	0.85	0.85
IT	1.07	1.07	1.07	1.06	1.06	1.06
BN	0.85	0.85	0.86	0.88	0.88	0.88
IB	0.93	0.93	0.94	0.94	0.94	0.94

NUC	1995	2000	2005	2010	2015	2020
FR	0.51	0.51	0.54	0.57	0.57	0.57
UK	0.81	0.81	0.82	0.83	0.83	0.83
IT	0.74	0.74	0.76	0.77	0.77	0.77
BN	0.73	0.73	0.74	0.76	0.76	0.76
IB	0.81	0.81	0.82	0.83	0.83	0.83

Fuel prices also differ between countries, due to distortions in competition or differences in transportation costs. Fuel prices in Germany are shown in *Table 5.6.* and relative fuel prices in the other regions in *Table 5.7.* The data for Germany were taken from PROGNOS (1995), and interregional differences were calculated on the basis of IEA (1996). The fuel price assumptions refer to the closed market scenario. In the open market case they are assumed to converge in the same way as investment costs.

Table 5.6. Fuel Prices in Germany (DM/GJ)

	1995	2000	2005	2010	2015	2020
HC	3.242	3.812	4.002	4.210	4.497	4.804
BC	3.390	3.310	3.183	3.303	3.479	3.655
OIL	4.560	5.643	5.783	6.390	6.855	7.424
GAS-CC	5.900	6.210	6.826	7.595	8.569	9.601
GASTUR	7.222	7.602	8.356	9.297	10.489	11.752
NUC	2.229	1.690	1.628	1.566	1.566	1.566

Table 5.7. Fuel Prices Relative to German Fuel Prices

HC	1995	2000	2005	2010	2015	2020
FR	1.00	1.00	1.00	1.00	1.00	1.00
UK	0.95	0.95	0.95	0.96	0.96	0.96
IT	1.32	1.10	1.00	1.00	1.00	1.00
BN	0.95	0.95	0.95	0.96	0.96	0.96
IB	1.08	1.08	1.07	1.06	1.06	1.06

Table 5.7. (continued)

OIL	1995	2000	2005	2010	2015	2020
FR	0.98	0.98	0.98	0.98	0.98	0.98
UK	0.70	0.70	0.72	0.74	0.74	0.74
IT	0.96	0.96	0.96	0.97	0.97	0.97
BN	0.92	0.92	0.93	0.93	0.93	0.93
IB	0.89	0.89	0.90	0.91	0.91	0.91

GAS-CC	1995	2000	2005	2010	2015	2020
FR	1.00	1.00	1.00	1.00	1.00	1.00
UK	0.94	0.90	0.86	0.88	0.88	0.88
IT	0.98	0.98	0.98	0.98	0.98	0.98
BN	0.95	0.95	0.95	0.96	0.96	0.96
IB	1.00	1.00	1.00	1.00	1.00	1.00

NUC	1995	2000	2005	2010	2015	2020
FR	0.63	0.63	0.66	0.68	0.68	0.68
UK	0.75	0.75	0.77	0.78	0.78	0.78
IT	1.00	1.00	1.00	1.00	1.00	1.00
BN	0.63	0.63	0.66	0.68	0.68	0.68
IB	0.63	0.63	0.66	0.68	0.68	0.68

The development of *labor costs* over time is displayed in *Table 5.8.* Since labor costs are a relatively insignificant cost category in comparison to capital and fuel costs, no attempt has been made to differentiate them with respect to region. Due

Table 5.8. Labor Costs (DM/kW)

	1995	2000	2005	2010	2015	2020
HC	15.0	14.9	14.7	14.6	14.4	14.3
HC-CC	15.0	14.9	14.7	14.6	14.4	14.3
HCGAS	13.3	13.2	13.0	12.9	12.8	12.6
BC	17.8	17.6	17.4	17.3	17.1	16.9
OIL	10.0	9.9	9.8	9.7	9.6	9.5
GAS-CC	6.0	5.9	5.9	5.8	5.8	5.7
GASTUR	2.7	2.7	2.6	2.6	2.6	2.6
NUC	15.0	14.9	14.7	14.6	14.4	14.3

Labor costs are expressed at 1985 price level.

to increasing automatization, labor cost decrease over time. Lignite-fired plants have the highest labor requirements per unit of capacity, whereas gas-fired ones have the lowest.

The *technical lifetime* of power stations is assumed to be 30 to 40 years, depending on plant type. There is no fixed economic lifetime. Rather, it is assumed that plant availability decreases and maintenance and repair costs increase towards the end of the technical lifetime, such that the actual decommissioning of plants is subject to an economic optimization calculus.

Of considerable importance for the effect of market opening are the *costs of electricity trade*. These costs consist of the costs of additions to transmission capacities, if necessary, and in transmission losses. While the effect of capacity additions on average electricity costs is small, transmission losses are a potentially important barrier to interregional electricity trade. Transmission losses are assumed to be 10 percent per 1000 km.

5.1.4 Exogenous Drivers

All simulations performed are based on common assumptions for exogenous variables. The exogenous driving variables of the model are the import volumes and the export prices of the rest of the world (ROW), and the rates of growth of electricity demand. The import volumes of the rest of the world are assumed to grow at 2.2 percent annually, while the export prices (relative to the numeraire) are assumed to be constant, except for the energy carriers. Electricity demand grows at different rates in the various regions. The growth of electricity demand has been taken from PROGNOS (1995) and UNIPEDE (1995).

The annual growth rates of the exogenous variables are compiled in *Table 5.9*.

Table 5.9. Annual Growth Rates of Exogenous Variables (Percent)

Electricity Demand

	95-00	00-05	05-10	10-15	15-20
GE	1.10	1.00	0.99	0.76	0.75
FR	1.70	1.41	1.28	1.20	1.14
UK	1.70	1.50	1.10	1.00	1.00
IT	2.30	2.10	1.40	1.20	1.10
BN	1.44	1.26	1.16	1.06	0.98
IB	3.67	2.78	2.33	2.00	1.80

Table 5.9 (continued)

Energy Export Prices of ROW

	95-00	00-05	05-10	10-15	15-20
HC	0.00	0.00	0.11	0.11	0.11
OIL	1.41	1.32	1.23	1.16	2.15
GAS	1.97	1.98	2.49	2.36	2.11

Import Volume of ROW

	95-00	00-05	05-10	10-15	15-20
ROW	2.20	2.20	2.20	2.20	2.20

Besides these exogenous variables, the various rates of technical progress determine the model dynamics. They are shown in *Table 5.10.*

Table 5.10. Annual Growth Rates of Factor Productivities 1995-2020 (Percent)

	GE	FR	UK	IT	BN	IB
Energy	3.00	2.00	2.70	2.00	2.50	1.00
Labor	1.50	1.50	1.50	1.50	1.50	1.50
Capital	1.50	1.50	1.50	1.50	1.50	1.50

5.2 Initial Economic Situation and Scenario Definitions

5.2.1 Initial Situation

The economic consequences of CO_2 abatement depend to a large extent on the structural characteristics of each particular economy.
Table 5.11. shows the percentage share in gross output of the various sectors in the six model regions.

Table 5.11. Sectoral Shares in Gross Output (Percent)

	GE	FR	UK	IT	BN	IB
Agriculture	2.20	5.12	2.83	4.02	4.61	6.47
Hard Coal	0.83	0.21	0.82	0.08	0.21	0.41
Lignite	0.09	0.00	0.00	0.00	0.00	0.11
Oil	2.19	3.16	6.14	3.33	3.30	4.57
Gas	0.95	0.75	1.20	0.47	2.33	0.15
Electricity	2.08	1.93	2.01	1.69	1.50	2.32
Basic Materials	10.38	7.15	6.81	8.83	9.42	9.32
Equipment	16.96	11.94	11.57	12.13	8.54	9.55
Consumption Goods	13.31	13.57	13.59	17.4	14.72	17.65
Construction	5.95	7.74	6.95	8.66	9.03	8.36
Transportation	3.25	3.82	4.12	4.26	4.92	4.50
Private Services	28.79	32.33	26.14	29.75	31.95	28.84
Public Services	13.00	12.27	17.82	9.38	9.47	7.76

It can be expected that the economic consequences of CO_2 abatement will be more pronounced in countries, in which the energy industries are important production sectors, than in others, in which they are not. As can be seen from Table 5.11, the importance of the energy sectors differs somewhat across regions. In Germany, France and Italy their share is about 6 percent; in Belgium/Netherlands and Spain/Portugal it is about 7.5 percent, and in the U.K. it is 10 percent. The greater importance of the energy industries in Spain/Portugal and especially in the U.K. is due to the large petroleum industries in these countries. In the case of Belgium/Netherlands the (Dutch) gas industry is particularly important.

Among the non-energy industries, agriculture is of minor importance in Germany and the U.K., whereas it takes a larger share in France, Spain and Portugal. Basic materials and equipment goods are more important industries in Germany than in any of the other regions, whereas consumption goods and, especially, construction are less important in Germany than elsewhere. The shares of services (including transportation and public services) are in a range of 40-50 percent, with France and the U.K. ranging near the top of the interval and Italy, Spain and Portugal near the bottom.[6]

It is obvious that the effect upon the various non-energy sectors of policies to reduce CO_2 emissions will depend on their energy intensity. In addition, the structure of demand (consumption, investment, export) has a strong influence on the way in which a particular sector is affected by CO_2 abatement policies. Especially, a sector's export share is usually considered an important

[6] In the case of the U.K. the high share of public services visible in the benchmark data has meanwhile diminished due to privatization.

characteristic, because CO_2 abatement is frequently expected to affect international competitiveness. Also, the ratio of consumption demand to investment demand is important, as these two demand categories may be affected differently by CO_2 abatement.

Table 5.12 shows the energy intensities (along with the labor and capital intensities) and the demand composition of the various non-energy sectors.

While the energy intensity is below 3 percent in most sectors, it is well above 10 percent in the basic materials and in the transportation sector. It is remarkable that the energy intensity of the transportation sector shows a considerable dispersion across regions. While it is only 10 percent in Germany, the corresponding figure is 16.5 percent in Spain/Portugal. Conversely, the agricultural sector displays a much higher energy intensity in Germany than in any other region.

Considering the demand composition, it is not surprising that Belgium and the Netherlands, being small countries, have the largest export share. This is true for all sectors. In Germany, France and Italy the export share is largest in the equipment goods industry. In the U.K. export dependence is largest in the basic materials industry, while in Belgium, the Netherlands, Spain and Portugal the transportation industry is the largest exporter. In all regions, investment demand is (unsurprisingly) the most important demand category for the construction and the equipment industry. It is also not surprising that consumption demand is most important for the consumption goods industry and for the private service sector.

Other important structural features influencing the development of CO_2 emissions are the CO_2 intensity of GDP and the fuel mix in power generation. The latter is important, because the structure of power generation capacities changes only gradually, depending on the lifetime of existing plants.

Table 5.12. Factor Intensities and Demand Composition (Percent)

GE	Labor Intensity [a]	Capital Intensity [b]	Energy Intensity [c]	Share of Intermediates [d]	Share of Consumption [d]	Share of Investment [d]	Share of Exports [d]
Agriculture	13.17	450.97	9.04	75.95	19.03	0.21	4.8
Basic Materials	17.98	95.28	10.33	70.36	4.23	0.06	25.35
Equipment	32.46	71.84	1.92	34.12	9.09	20.51	36.28
Consumption Goods	19.57	76.43	2.93	43.79	40.56	1.28	14.36
Construction	33.04	41.4	2.29	21.68	7.35	67.41	3.56
Transportation	32.25	336	10.14	57.9	17.08	0.7	24.33
Private Services	23.82	409.41	1.87	51.76	42.12	3.42	2.7
Public Services	48.06	221.12	2.47	16.17	3.31	0	0.13

FR	Labor Intensity [a]	Capital Intensity [b]	Energy Intensity [c]	Share of Intermediates [d]	Share of Consumption [d]	Share of Investment [d]	Share of Exports [d]
Agriculture	8.68	473.34	3.29	67.08	18.66	0.56	13.7
Basic Materials	21.13	102.37	12.1	64.25	8.24	0	27.51
Equipment	29.72	67.75	1.73	36.31	11.65	23.88	28.16
Consumption Goods	22.07	81	2.54	42.41	42.15	0.57	14.87
Construction	30.54	43.75	2.01	17.22	10.75	70.57	1.47
Transportation	40.07	353.99	12.6	62.45	21.37	0	16.18
Private Services	31.99	456.75	1.59	42.95	50.28	2.39	4.38
Public Services	63.13	205.76	2.67	0	7.57	0	0

Table 5.12. (continued)

UK	Labor Intensity [a]	Capital Intensity [b]	Energy Intensity [c]	Share of Intermediates [d]	Share of Consumption [d]	Share of Investment [d]	Share of Exports [d]
Agriculture	10.82	392.68	5.67	74.76	18.87	0	6.37
Basic Materials	18.78	94.51	11.81	67.14	5.52	0.02	27.32
Equipment	27.86	63.25	1.94	44.09	8.35	22.54	25.02
Consumption Goods	18.31	58.43	2.07	48.35	37.34	2.54	11.77
Construction	18.45	35.98	2.16	36.46	4.31	58.61	0.62
Transportation	38.85	354.26	12.92	55.47	22.24	2.9	19.39
Private Services	30.72	347.25	2.83	47.31	40.8	2.56	9.33
Public Services	61.81	216.32	2.44	7.27	30.7	0	1

IT	Labor Intensity [a]	Capital Intensity [b]	Energy Intensity [c]	Share of Intermediates [d]	Share of Consumption [d]	Share of Investment [d]	Share of Exports [d]
Agriculture	17.35	414.74	3.87	69.6	25.13	0.32	4.96
Basic Materials	15.91	94.01	9.77	74.03	9.53	1.02	15.41
Equipment	24.37	72.38	1.93	35.43	11.32	27.42	25.83
Consumption Goods	17.21	81.02	2.61	42.27	39.88	1.11	16.74
Construction	20.21	41.81	1.39	25.99	10.01	63.89	0.11
Transportation	36.07	339.12	15.15	53.71	23.83	2.78	19.69
Private Services	21.18	413.23	2.63	41.4	52.24	2.48	3.89
Public Services	71.56	223.11	2.59	0.02	4.1	0	0.12

Table 5.12. (continued)

BN	Labor Intensity [a]	Capital Intensity [b]	Energy Intensity [c]	Share of Intermediates [d]	Share of Consumption [d]	Share of Investment [d]	Share of Exports [d]
Agriculture	6.54	320.61	4.48	64.37	10.88	0	24.74
Basic Materials	18.72	104.41	11.53	45.5	3.69	0.25	50.55
Equipment	28.34	66.23	1.53	31.51	8.55	18.53	41.4
Consumption Goods	17.27	67.82	1.97	37.15	26.06	1.04	35.75
Construction	26.36	35.97	1.77	33.13	10.11	52.07	4.68
Transportation	39.14	299.11	11.08	33.83	13.83	0.69	51.65
Private Services	33.45	331.56	1.79	36.43	50.06	3.41	10.1
Public Services	63.92	171.69	1.99	7	2.56	0.71	0.37

IB	Labor Intensity [a]	Capital Intensity [b]	Energy Intensity [c]	Share of Intermediates [d]	Share of Consumption [d]	Share of Investment [d]	Share of Exports [d]
Agriculture	13.93	373.7	4.89	76.15	15.78	0.45	7.62
Basic Materials	18.01	92.87	10.75	72.12	7.12	0.21	20.55
Equipment	28.35	71.14	1.81	40.12	12.48	22.39	25.01
Consumption Goods	17.37	69.9	2.65	39.61	41.54	1.89	16.97
Construction	23.36	36.08	3.03	23.43	14.43	61.89	0.25
Transportation	34.78	353.93	16.54	44.61	25.12	1.13	29.14
Private Services	20.83	363.87	2.72	36.17	57.31	3.32	3.2
Public Services	66.46	153.87	3	0	7.09	0	0

[a] Share of wages in gross output.
[b] Ratio of capital stock to gross output
[c] Share of energy costs in gross output
[d] Share of demand category in total demand

Initial CO_2 intensities of GDP are displayed in *Table 5.13*. They are highest in the U.K., in Belgium/Netherlands and Spain/Portugal, and relatively low in France and Italy. Germany assumes an intermediate position. The reason for the low value in Italy may be partly meteorological (relatively low heating requirements). Similar reasoning may apply to the southern part of France, but the main reason for the low CO_2 intensity in France is the high share of nuclear power (see below). The high figure for Spain/Portugal is somewhat unexpected in view of meteorological conditions. It indicates low energy efficiency; in part it is due to the relatively high share of coal in the energy mix.

Table 5.13. CO_2 Intensities of GDP (t CO_2/1000 ECU)

GE	FR	UK	IT	BN	IB
0.78	0.52	0.87	0.63	0.95	0.92

To provide a closer look at the electricity industry, *Table 5.14.* shows the structure of power generation by energy carrier.

Table 5.14. Structure of Power Generation in the Base Year (Percent)

	GE	FR	UK	IT	BN	IB
Hard Coal	23.70	4.81	43.74	8.59	31.79	34.50
Lignite	30.23	0.00	0.00	0.58	0.00	6.95
Oil	0.75	0.95	10.97	61.63	0.26	7.97
Gas	7.07	0.84	15.96	8.56	32.18	0.35
Nuclear	33.81	77.51	27.56	0.00	33.87	29.93
Others	4.43	15.89	1.77	20.65	1.91	20.3

It can be seen that the unique characteristic of the German power generation industry is the strong dependence on brown coal. Important features of the other national electricity supply systems are the high share of nuclear power in France and the absence of nuclear power in Italy, combined with a strong dependence on oil. Gas plays a significant role primarily in the U.K. and in Belgium/Netherlands (which in this case means: the Netherlands).

From the point of view of CO_2 emissions the most important structural feature of the national electricity supply systems is the share of fossil fuel based generation. With respect to this feature, a rather substantial dispersion is noteable. Fossil shares range from 7 percent in France to 79 percent in Italy.

5.2.2 Policy Scenarios

Our policy simulations consider the following scenarios:

1. *Base Case* - 'Closed' national electricity markets are present: electricity imports are bounded by current import shares. No specific CO_2 abatement policies are pursued.
2. *Internal Electricity Market* - There is an internal European electricity market without import restrictions. No specific CO_2 abatement policies are pursued.
3. *Internal Electricity Market cum German Nuclear Phase-Out* - An internal European electricity market without import restrictions is established. Germany dismantles its nuclear capacity by 2005. No specific CO_2 abatement policies are pursued.
4. *Internal Electricity Market cum European Carbon/Energy Tax* - A European carbon/energy tax is enacted in combination with a European internal electricity market.
5. *Internal Electricity Market cum German Carbon/Energy Tax* - A unilateral carbon/energy tax in Germany is enacted in combination with a European internal electricity market.
6. *National Tradeable Permit Markets* - A European internal electricity market is established. Uniform carbon reduction targets across countries are implemented using national tradeable permits
7. *European Tradeable Permit Market* - A European internal electricity market is established. An overall European CO_2 reduction target which is equivalent to the overall reduction effect of the national abatement targets is implemented using European-wide tradeable emission permits.

Part II: Simulation Results

6 Effects of Market Opening and National Energy Policies

6.1 Introduction

This chapter considers the economic and environmental (in terms of CO_2) effects of creating a European Internal Market for electricity. Two scenarios with respect to nuclear power utilization in Germany will be applied. The first assumes that nuclear capacities in Germany remain at their current level. The second assumes a gradual phase-out until 2005. In both cases it is assumed that no specific measures to reduce CO_2 emissions will be undertaken.

The effect of market opening will be expressed in terms of changes from a baseline with 'closed' national electricity markets. 'Closed' means that each region's share of electricity imports from total demand will be restricted not to exceed its current level. In contrast to this scenario, the Council of Energy Ministers recently passed a directive stating that electricity customers above certain thresholds of electricity demand will be free to purchase electricity from foreign suppliers. More specifically, a timetable has been agreed to, in which free choice of supplier is to be established by 1999 for customers with a demand larger than 40 GWh, and by 2005 for customers with a demand larger than 9 GWh.

Based on the structure of electricity demand, these provisions can be translated into certain percentages of each national market opening to foreign suppliers. For 1999 and 2005 this would entail approximate maximum import shares of 23 percent and 33 percent, respectively.

In contrast to this schedule, the Internal Market scenario used in this study will be based on the assumption that import restrictions are *entirely* abolished. It will then be interesting to see how imports resulting from such an unrestricted scenario compare to the limits specified in the directive.

In running these simulations it is assumed that competitive pressure will lead to a convergence of capital and fuel costs currently differing across regions (see section 5.1.3). The economic reasoning behind this is that price differentiation and overly demanding technical specifications - which may occur in separated markets - cannot be sustained in an open market with arbitrage possibilities. In addition, it is assumed that the discount rate utilized in the investment calculus is higher in the open market case than in the closed market case, reflecting a premium for increased market risk.

To provide a benchmark for the Internal Market scenarios we will first consider the base case, in which import shares are limited to their current levels, cost differences remain, and discount rates are low.

6.2 The Base Case

The average annual percentage changes of CO_2 and GDP are displayed in *Table 6.1*. GDP growth rates are relatively uniform across regions, ranging from slightly less than 2 percent in France to almost 2.3 percent in Italy. The growth rates of CO_2 show substantially more dispersion, ranging from 0.8 percent in Germany to almost 1.7 percent in the U.K. The CO_2 intensities of GDP decrease in all regions, but to varying degrees. The reduction in CO_2 intensity is most pronounced in Germany and smallest in the U.K.

Table 6.1. Average Annual Change of CO_2 and GDP, 1995-2020 (Percent)

	GE	FR	UK	IT	BN	IB
CO_2	0.81	0.97	1.67	1.06	1.55	1.01
GDP	2.16	1.95	2.17	2.27	2.12	2.04
CO_2/GDP	-1.35	-0.98	-0.50	-1.21	-0.57	-1.03

There are several factors that contribute to these developments of CO_2 intensities in the various regions:

- *The development of overall energy prices.* A stronger increase in energy prices leads to a greater reduction in fossil energy and, hence, CO_2 intensity, other things equal.
- *The initial CO_2 intensity.* The effect of fossil energy price increases on energy intensity and, hence, CO_2 intensity is large if the initial intensity is high.[1]
- *The development of the energy mix.* If the prices of the various fuels develop differently, the energy mix will change in favor of more CO_2 intensive (coal) or less CO_2 intensive (gas) fuels.

These factors will now be considered in an effort to explain the changes in CO_2 intensities described above.

Table 6.2 shows the average annual changes of CO_2 emissions by energy carrier, and *Table 6.3* displays the corresponding price changes.

[1] Technically speaking: due to the concavity of the production function, the reduction in energy input induced by a given price increase is large, if the initial marginal product of energy is low.

Table 6.2. Average Annual Change of CO_2 by Energy Carrier, 1995-2020 (Percent)

	GE	FR	UK	IT	BN	IB
HC	0.25	1.75	0.43	4.87	1.73	0.73
BC	1.19	0.00	0.00	-9.28	0.00	1.05
OIL	0.38	0.52	1.02	-0.52	0.36	0.74
GAS	2.03	1.68	2.97	2.49	2.30	4.39

Table 6.3. Average Change of Energy Prices, 1995-2020 (Percent)

	GE	FR	UK	IT	BN	IB
HC	1.58	1.39	0.79	1.35	1.40	1.14
BC	-0.14	0.00	0.00	0.00	0.00	-0.35
OIL	1.41	1.20	0.17	1.20	1.42	1.37
GAS	-0.20	0.51	-0.84	1.22	-0.25	7.63

Considering Germany, we find that, except for gas, the energy prices increase relatively strongly in comparison with the other regions. This leads to a rather weak increase in energy demand relative to GDP. Also, the energy mix shifts in favor of gas (due to the price reduction of gas). Together, these factors explain the strong fall in the CO_2 intensity of German GDP.

For France, we find a more moderate increase in the energy price level. Together with the low initial fossil energy intensity this implies that fossil energy demand is reduced less than in Germany. In addition, hard coal utilization increases somewhat stronger than that of gas. For these reasons, the CO_2 intensity decreases less in France than in Germany.

The U.K. is characterized by the smallest increase (coal, petroleum) or strongest decrease (gas) in energy prices across the EU. This explains why the reduction in CO_2 intensity is very low in the U.K.

For Italy, the reduction in energy intensity is quite strong. This results from the increase in the energy price level, which dominates the effect of the increased share of hard coal in the fuel mix.

In Belgium/Netherlands, the increase in the overall energy price is moderate (if one considers the decrease in the gas price in combination with the high share of gas in the fuel mix). Fossil energy input increases quite considerably. This implies a relatively small reduction in CO_2 intensity.

In Spain/Portugal we find a moderate increase in the coal price and a strong increase in the gas price. Given the high initial CO_2 intensity, the overall price increase implies a considerable reduction in the CO_2 intensity.

With respect to the electricity supply industry, it is instructive to compare the expected future supply structure with the present structure as described in the preceding section.

Table 6.4 displays the supply structure by 2020. A trivial difference is that the share of nuclear power is in general smaller than at present. This is due to our assumption that the expansion of nuclear capacities (in absolute terms) is bounded by assumption, whereas overall electricity demand increases (see section 5.1.4). Consequently (since the potential for expanding hydro-electric power is limited) the share of fossil fuel based electricity increases. The most important changes to take place are the increase in the coal share in France, Italy and, to a lesser extent, in Belgium/Netherlands, and the increase in the gas share in Italy, Spain/Portugal and, to a lesser extent, in the U.K. Oil is reduced to negligible importance except in the U.K. and Italy.

Table 6.4. Structure or Power Generation in 2020 (Percent)

	GE	FR	UK	IT	BN	IB
HC	26.24	8.70	43.15	43.22	45.43	25.62
BC	32.28	0.00	0.00	0.04	0.00	5.65
OIL	0.03	0.06	9.34	7.60	0.78	5.02
GAS	9.39	1.48	23.43	33.11	26.57	29.99
NUC	27.41	77.73	22.45	0.00	25.92	18.65
Others	4.66	12.03	1.63	16.03	1.30	15.07

The development of the CO_2 emissions stemming from the electricity sector must also be examined. *Table 6.5* shows the average annual percentage rates of change over the period 1995-2020. These rates of change are, of course, a reflection of the changes in the supply structure just described. Consequently, the emissions from oil utilization in power generation decrease in most regions, whereas emissions from gas rise substantially in most cases (particularly in Italy and Spain/Portugal). Coal-based electricity-related emissions rise sharply in France and Italy.

Table 6.5. Average Growth Rate of Electricity-Related CO_2 Emissions 1995-2020 (Percent)

	GE	FR	UK	IT	BN	IB
HC	0.61	3.59	0.39	8.17	2.49	0.46
BC	0.44	0.00	0.00	-9.39	0.00	0.59
OIL	-11.36	-9.57	0.28	-6.88	6.06	0.13
GAS	0.34	2.02	2.27	5.22	0.07	19.21

This concludes the description of the base case ('closed market') scenario.

6.3 Internal Market for Electricity

The discussion of the effects of creating an Internal Market for electricity starts with the first-order effects, i.e. those on the electricity supply industry.[2] The impact on other industries and on the macroeconomy is a derived one and will be addressed subsequently.

Theoretically, the internal electricity market may have several (partial) effects on electricity-related CO_2 emissions. A *level* effect may arise, if the opening of markets leads to a change in overall electricity output. This study proceeds with the assumption that electricity demand is exogenously given. The effect on CO_2 is, thus, an entirely *structural* one, relating to changes in the location of power generation and in the fuel and technology mix.[3]

The effect upon the European electricity industry of an internal electricity market is presented in *Table 6.6* in terms of percentage changes from the 'closed market' case. Obviously, a substantial relocation of total power generation to the various countries occurs. In comparison with the 'closed market' case, power generation decreases in Germany, Italy and Spain/Portugal, and increases in France, the U.K. and Belgium/Netherlands. The *increase* in France and the U.K. is more pronounced in the intermediate term (2005) than in the long term (2020), whereas for the increase in Belgium/Netherlands the reverse is true. On the other

Table 6.6. Effect of Market Opening on Power Generation, Trade and Electricity-Related CO_2 Emissions (Percent)

2005	GE	FR	UK	IT	BN	IB	EU
Total Gen.	-0.88	7.23	7.28	-19.13	1.62	-2.84	0.12
HC	-4.35	114.68	10.80	-0.62	-4.91	-3.13	6.48
BC	0.00	0.00	0.00	29.13	0.00	0.00	0.02
Oil	179.44	0.00	33.16	-13.61	0.00	0.00	-1.45
GAS	10.15	133.75	4.69	-72.86	23.85	4.15	-7.98
NUC	0.00	0.00	0.00	0.00	-7.17	-8.99	-1.09
Others	0.00	2.60	0.00	1.81	0.00	-4.80	-0.07
Imports	-2.21	896.39	-100.00	148.32	-1.24	151.91	62.12
Exports	-13.73	75.74	0.00	0.00	45.99	9.04	64.74
CO_2.	-1.68	133.67	13.51	-14.72	0.31	-1.98	2.76

[2] See also Hoster (1996b).

[3] Minor changes in overall power generation arise as a result of changes in transmission losses induced by changes in the level of trade. This applies to all policy simulations conducted.

Table 6.6. (continued)

2020	GE	FR	UK	IT	BN	IB	EU
Total Gen.	-4.74	2.71	2.50	-1.23	11.40	-1.30	0.05
HC	-19.81	17.80	-8.58	-38.91	-50.49	-24.30	-21.93
BC	-18.83	0.00	0.00	0.00	0.00	0.00	-16.81
Oil	3411.11	873.08	0.00	103.96	0.00	0.00	42.21
GAS	55.81	2.23	26.67	22.65	129.50	16.39	33.45
NUC	0.00	0.00	0.00	0.00	0.00	0.00	0.00
Others	0.00	5.29	0.00	1.78	0.00	2.71	2.30
Imports	49.22	220.95	-57.63	17.98	11.74	84.51	34.51
Exports	-6.61	27.33	0.00	2847.78	342.47	79.41	35.38
CO_2	-11.83	26.98	-1.30	-9.51	-10.75	-7.54	-6.89

hand, the *reduction* in German electricity generation is substantially stronger in the long term than in the intermediate term, whereas the opposite is true for the reduction in Italy and Spain/Portugal. It can be seen that in the long term most of the regions increase both their exports *and* imports. Exceptions are Germany, where one observes a slight reduction of exports, and the U.K., where imports decrease.

Overall, power generation in Germany is permanently, and to a considerable degree, substituted by imports. In Italy, Spain and Portugal there is a substitution of imports for domestic generation as well, but the induced long-term reduction of power generation is less pronounced than in Germany.

Comparing the electricity imports arising in the unrestricted-trade scenario with the import limits implicit in the European Commission's Internal Market directive (see section 6.1), it can be seen from *Table 6.7* that the import shares stay well below those limits. Only for Italy the import share in 2005 comes close to the limit.

Table 6.7. Electricity Imports as a Percentage of Domestic Demand in the Internal Market Scenario

	GE	FR	UK	IT	BN	IB
Limit 2005	33.0	33.0	33.0	33.0	33.0	33.0
Unrestricted 2005	7.4	3.9	0.0	28.4	12.0	4.7
Unrestricted 2020	12.8	1.5	1.1	11.8	6.9	3.0

An important message conveyed by Table 6.7 is that import shares do not necessarily rise over the course of time. With the convergence of costs induced by the Internal Market, the incentive to import electricity tends to decrease, and import shares converge to levels that are sustainable in view of persistent cost differences.

With respect to CO_2, it can be seen that in the intermediate term any increase in a country's power generation goes along with increased CO_2 (France, U.K., Belgium/Netherland), whereas decreasing generation implies decreasing CO_2 emissions. In France and the U.K. the percentage increase in CO_2 is larger than the increase in power generation, because additional generation is largely coal based. In Belgium/Netherlands a strong increase in gas based electricity occurs, leading to a decrease in the CO_2 intensity of power generation. Overall, however, the gas share is reduced and the coal share is increased, implying an increase in European electricity-related CO_2 emissions by 2.8 percent.[4]

In the long term, the situation is different. There is a strong long-term substitution of gas for coal in all regions except for France. Thus, the Internal Market leads to an increase in long-term emissions only in France. In all other regions CO_2 emissions decrease in the long term, even if a substantial increase in power generation is registered. Overall, European electricity-related CO_2 emissions decrease 6.9 percent by 2020.

The Internal Electricity Market visibly affects several other sectors of the economy and, ultimately, the macroeconomy. These effects are presented in *Table 6.8* and *Table 6.9*, respectively.

Table 6.8. Sectoral Effects of Market Opening (Percent)

2005	GE	FR	UK	IT	BN	IB	EU
Agriculture	0.03	0.15	0.31	0.07	0.08	0.01	0.11
Hard Coal	0.00	17.59	5.83	10.12	-5.12	-3.32	2.78
Lignite	0.06	0.00	0.00	0.00	0.00	0.00	0.04
Oil	1.83	0.83	-0.62	2.12	3.05	2.03	0.82
Gas	0.39	0.42	0.72	-11.85	-0.04	1.08	-0.68
Electricity	-0.88	7.23	7.28	-19.13	1.62	-2.84	0.12
Basic Materials	-0.15	0.33	0.31	0.38	0.02	-0.27	0.09
Equipment	-0.24	0.01	0.34	-0.28	-0.18	-0.21	-0.10
Consumption G.	-0.04	0.19	0.33	0.18	0.09	-0.03	0.12
Construction	0.02	0.36	0.71	-0.25	-0.21	0.10	0.15
Transport	-0.02	0.33	0.30	0.09	0.05	0.11	0.15

[4] It can be added that the increase in electricity-related CO_2 is partly due to a reduction in nuclear power generation in Belgium/Netherlands and in Spain/Portugal. This is an effect of the increased interest rate in the open-market case. However, the higher interest rate does not imply a *permanent* reduction of nuclear-power utilization in these two regions. Rather, the building of nuclear capacity is *postponed*. By 2010 and beyond, nuclear power generation is identical in the closed market case and in the open market case, and the increase in CO_2 is converted into a reduction (see below).

Table 6.8. (continued)

Private Services	-0.01	0.29	0.30	0.10	0.07	0.04	0.14
Public Services	-0.03	0.17	0.29	-0.02	0.05	-0.12	0.10

2020	GE	FR	UK	IT	BN	IB	EU
Agriculture	0.15	0.09	0.31	0.21	-0.20	-0.28	0.08
Hard Coal	0.00	-1.52	-6.70	-26.84	-20.84	-8.33	-9.91
Lignite	-8.33	0.00	0.00	0.00	0.00	-0.34	-6.57
Oil	5.32	2.46	-1.68	6.51	6.17	6.87	1.85
Gas	3.47	0.86	1.79	4.58	1.45	9.26	2.57
Electricity	-4.74	2.71	2.50	-1.23	11.40	-1.30	0.05
Basic Materials	0.04	-0.27	0.11	0.06	-0.72	0.34	-0.03
Equipment	-0.14	-0.61	0.05	-0.04	-0.96	0.71	-0.16
Consumption G.	0.15	-0.05	0.21	0.32	-0.30	-0.45	0.07
Construction	-0.04	0.28	0.66	-0.22	-0.13	-1.01	0.04
Transport	0.15	0.16	0.15	0.20	-0.31	0.07	0.11
Private Services	0.21	0.31	0.29	0.27	0.40	-1.57	0.14
Public Services	0.12	0.12	0.05	0.32	0.48	-1.59	0.07

The effects on the fuel industries are basically a mirror image of the changes in fuel use in the electricity sector. Especially, there is a long-term substitution of gas and oil for hard coal and lignite.

With respect to the non-energy industries, it can be seen that the reduction of electricity generation in Germany leads to a slight decline in Germany's basic material and equipment goods industries, at least in the intermediate term. Similar intermediate-term tendencies are visible in Spain/Portugal and, to a lesser extent in Italy. In the long term, when the induced reduction in electricity generation becomes smaller, these effects tend to be reversed. The effect upon the electricity exporting countries, France and the U.K., is expansionary for most sectors, in the intermediate term. In the long term, the effect reverses for the French materials and equipment industry. For Belgium/Netherlands the effect is mixed, both in the intermediate and in the long term.

Turning to the general economy (*Table 6.9*), it is found that the effect on GDP approximately corresponds to that on electricity generation. In the long term, France, the U.K. and Belgium/Netherlands experience a slight increase, and Germany and Spain/Portugal a decrease in GDP. However, the long-term effect is pronounced only in the case of Spain/Portugal. In the intermediate term, the pattern of effects on GDP differs somewhat from that on electricity generation. Italy experiences a small increase in GDP, and the GDP of Belgium/Netherlands is almost unaffected by the internal electricity market.

The effect on employment is rather small on an EU-wide scale and tends to be positive. In the long term, this holds for each individual country as well. In the intermediate term, there is a visible increase in France and the U.K. and a similar decrease in Italy.

Table 6.9. Effects of Market Opening on CO_2 and the Macroeconomy (Percent)

2005	GE	FR	UK	IT	BN	IB	EU
CO_2	0.47	8.26	3.39	1.89	0.98	0.57	2.49
Hard Coal	-2.28	46.00	8.42	-2.72	-3.13	-2.11	3.95
Lignite	0.07	0.00	0.00	32.70	0.00	0.03	0.10
Oil	2.44	1.83	3.95	0.62	2.68	2.27	2.23
Gas	0.35	1.62	1.07	10.97	1.89	-1.15	2.16
GDP	-0.12	0.24	0.27	0.15	-0.01	-0.20	0.08
Consumption	0.16	0.51	0.16	0.28	0.26	0.36	0.28
Investment	0.16	0.42	0.96	-1.04	-0.54	0.46	0.15
Export	-0.12	0.51	-0.71	0.39	0.17	-0.12	-0.01
Import	-0.22	0.32	0.00	-0.42	-0.10	-0.11	-0.08
Employment	-0.10	0.28	0.35	-0.35	-0.03	-0.10	0.04
Capital Stock	-0.02	-0.03	0.06	-0.02	-0.03	-0.06	-0.01
Exchange Rate	0.03	0.03	0.03	0.03	0.03	0.03	0.03

2020	GE	FR	UK	IT	BN	IB	EU
CO_2	-0.99	4.58	0.95	-2.67	-2.87	-1.23	-0.21
Hard Coal	-7.85	10.65	-6.04	-25.39	-29.33	-13.18	-11.73
Lignite	-10.03	0.00	0.00	0.00	0.00	-0.98	-8.95
Oil	5.17	4.22	2.49	11.17	4.88	3.36	5.25
Gas	0.88	1.21	2.89	-4.90	5.00	6.38	1.92
GDP	-0.12	0.05	0.10	-0.04	0.21	-1.80	-0.13
Consumption	0.38	1.02	0.42	0.49	1.55	-5.14	0.24
Investment	-0.21	0.78	0.96	-0.97	-0.26	-1.47	0.04
Export	-0.13	-0.44	-1.41	0.36	0.21	2.34	-0.13
Import	0.14	0.08	-0.44	-0.04	-0.03	0.44	-0.03
Employment	0.02	0.03	0.03	0.08	0.04	0.07	0.04
Capital Stock	0.01	0.09	0.35	0.10	-0.07	-0.12	0.10
Exchange Rate	-0.24	-0.24	-0.24	-0.24	-0.24	-0.24	-0.24

The intermediate-term CO_2 emissions increase in all countries. The overall increase is measured at about 2.5 percent. In the long term, a decrease in Germany, Italy, Belgium/Netherlands and Spain/Portugal can be seen and an increase in France and the U.K. The overall effect is a decrease of about 0.2 percent decrease.

Given that the imports arising from the unrestricted scenario are compatible with the European Commission's directive on market opening, we will take the unrestricted scenario as a description of the Internal Market. The effects of national energy policy, as well as those of CO_2 abatement policies, will be considered with this background.

6.4 Nuclear Phase-Out in Germany

This section considers the case of a German nuclear phase-out in the context of the Internal Market for electricity. More specifically, it is assumed that German nuclear capacities will be completely dismantled by 2005.

The effects of the German nuclear phase-out on the European electricity supply industry are displayed in *Table 6.10*, taking the Internal Market scenario with constant nuclear capacities in Germany as the standard of reference.

Table 6.10. Effect of German Nuclear Phase-Out on Power Generation, Trade and Electricity-Related CO_2 Emissions (Percent)

2005	GE	FR	UK	IT	BN	IB	EU
Total Gen.	-17.22	3.28	0.54	12.54	18.18	1.45	0.02
HC	11.32	3.62	0.28	-2.15	1.75	-1.76	3.41
BC	-0.05	0.00	0.00	0.00	0.00	0.00	-0.04
Oil	644.93	1181.79	0.00	-3.05	0.00	0.00	5.50
GAS	241.10	718.42	1.22	206.46	58.94	16.53	46.93
NUC	-100.00	0.00	0.00	0.00	0.00	0.00	-18.52
Others	0.00	0.46	0.00	-2.49	0.00	0.53	0.13
Imports	187.63	11.53	0.00	-30.69	11.79	-18.52	12.93
Exports	-49.65	16.52	14.12	1494.74	437.82	184.82	13.26
CO_2	11.01	16.36	0.56	5.99	15.88	0.36	7.77

2020	GE	FR	UK	IT	BN	IB	EU
Total Gen.	-13.45	0.69	1.75	4.43	14.30	1.14	0.02
HC	14.36	0.82	0.00	11.45	0.00	3.05	5.23
BC	23.20	0.00	0.00	0.00	0.00	0.00	20.29
Oil	0.00	13.23	0.00	2.65	0.00	0.00	1.69
GAS	39.27	31.80	6.11	2.66	26.60	1.55	14.75
NUC	-100.00	0.00	0.00	0.00	0.00	0.00	-16.83
Others	0.00	-0.16	0.00	-0.31	0.00	-0.77	-0.09
Imports	83.46	86.04	-55.74	-26.94	-7.84	-32.93	19.79
Exports	-31.94	10.97	116.63	104.60	98.00	48.22	20.16
CO_2	19.59	3.56	1.35	6.19	13.67	1.76	9.30

In Germany, the gap arising from nuclear phase-out amounts to 31.3 percent of total generation in the intermediate term (2005) and 27.4 percent in the long term (2020). As a result, German power generation is reduced by 17.2 percent in the intermediate term and 13.5 percent in the long term, the reduction being compensated by increased imports. The remaining gap is filled by increased fossil-fuel based electricity generation.

Because of increased German imports, power generation in all other regions rises. The increased power generation outside of Germany can be separated into a direct and an indirect effect. The direct effect arises in countries that satisfy increased German import demand (France, Belgium/Netherlands). The indirect effect arises in countries that need to compensate imports now absorbed by Germany (Italy, Spain/Portugal).

Considering the effect on power generation, CO_2, and capacities in more detail, the following is found.

In Germany a substantial change in the fossil fuel mix takes place. While hard-coal based power generation increases by 11.3 percent in the intermediate term, gas-based generation rises by more than 240 percent.[5] This change in the fuel mix limits the effect on CO_2 otherwise expected, so that the increase in electricity-related CO_2 amounts to about 11 percent.

Outside Germany one observes a similar development, i.e., the increased German import demand is mainly met by gas-based electricity.power.

Overall, the German nuclear phase-out initiates an increase in coal-based electricity throughout the EU of 3.4 percent and an increase in gas-based electricity of almost 47 percent. European CO_2 emissions rise by 7.8 percent.

A particularly effect to observe is the increase in hard coal-based electricity (especially in Germany) without necessitating capacity additions. The higher output level is entirely achieved by increasing plant utilization rates, which were in operation anyway. This is different for gas, where the rise in output requires corresponding capacity additions.

The long-term effect of the German nuclear phase-out differs somewhat from the effect in the intermediate term. Total German power generation is reduced less than it is in the intermediate term, i.e., a smaller increase in imports is recorded. The reason for this is that the convergence of costs implied by the Internal Market makes power generation in Germany more attractive in the long term. Thus, a substantial rise in electricity generation from German lignite occurs.

Due to smaller imports and a more unfavorable fuel mix (in terms of carbon content), German electricity-related CO_2 emissions now rise by more than 19 percent. Outside Germany, the German nuclear phase-out still leads to increased gas-based electricity generation (especially in France and Belgium/Netherlands).

---

[5] The percentage increase in oil-based electricity may appear even more impressive, but the share in oil based electricity still remains well below one percent.

Total CO_2 emissions increase by 9.3 percent. Capacity additions now occur mainly in Germany in the hard coal, gas and, especially, lignite branches.

The effects of the German nuclear phase-out upon the other sectors of the economy in Germany and abroad are displayed in *Table 6.11*.

Table 6.11. Sectoral Effects of German Nuclear Phase-Out (Percent)

2005	GE	FR	UK	IT	BN	IB	EU
Agriculture	-0.11	0.03	0.08	-0.10	-0.25	-0.01	-0.04
Hard Coal	0.00	2.53	0.76	-9.39	-3.53	-0.42	-0.92
Lignite	-0.41	0.00	0.00	0.00	0.00	0.00	-0.31
Oil	2.16	0.40	0.10	-0.56	-0.14	0.32	0.33
Gas	6.60	0.64	-0.01	10.51	0.86	8.41	2.87
Electricity	-17.22	3.28	0.54	12.54	18.18	1.45	0.02
Basic Materials	-0.46	0.33	0.23	-0.27	-0.81	0.09	-0.19
Equipment	0.66	-0.02	0.13	-0.06	-0.86	0.05	0.21
Consumption G.	0.15	0.01	0.09	-0.20	-0.57	-0.04	-0.04
Construction	-0.15	-0.33	-0.08	0.08	0.34	-0.10	-0.08
Transport	-0.09	0.00	0.06	-0.14	-0.40	-0.03	-0.08
Private Services	-0.17	-0.08	0.02	-0.14	0.14	-0.13	-0.08
Public Services	-0.06	0.01	0.04	-0.17	0.20	-0.08	-0.02

2020	GE	FR	UK	IT	BN	IB	EU
Agriculture	-0.05	0.03	0.03	-0.08	-0.26	0.04	-0.03
Hard Coal	0.00	0.99	0.09	3.86	-2.03	1.08	0.78
Lignite	8.83	0.00	0.00	0.00	0.00	0.00	6.75
Oil	0.73	0.02	0.06	-0.25	-0.05	0.19	0.10
Gas	4.26	0.16	0.15	1.16	0.47	1.40	1.32
Electricity	-13.45	0.69	1.75	4.43	14.30	1.14	0.02
Basic Materials	-0.21	0.00	-0.01	0.04	-0.68	0.25	-0.10
Equipment	0.36	-0.14	-0.15	-0.20	-0.80	0.10	0.01
Consumption G.	0.24	-0.02	-0.03	-0.08	-0.39	0.00	0.01
Construction	-0.29	0.00	0.09	-0.16	-0.06	-0.13	-0.09
Transport	-0.03	0.00	0.00	-0.09	-0.30	0.04	-0.05
Private Services	-0.14	0.01	0.02	-0.11	0.12	-0.19	-0.06
Public Services	0.03	0.00	-0.01	-0.20	0.26	-0.39	-0.02

By assumption, German hard coal mining is unaffected by the nuclear phase-out. Consequently, the German oil and gas industries expand strongly. In the long term, the output of the German lignite sector rises. On the European level, there is a long-term expansion of all fossil-fuel industries.

As concerns the German non-energy sectors, most of them experience a reduction in output. An exception is, unsurprisingly, the equipment goods industry, which experiences a visible expansion. This is true both in the intermediate and in the long term. In the long term a sharp increase in the lignite industry is triggered.

Outside Germany, the sectoral effects tend to differ from region to region, except for the electricity and gas industries, which experience an obvious increase both in the short and the long term.

Considering the environmental and macroeconomic effects, as shown in *Table 6.12*, one first observes that overall German CO_2 emissions increase by 2.3 percent (2005) and 3.5 percent (2020). The increase in emissions is mainly due to hard coal and lignite, and not gas, since the increase in gas utilization in the electricity industry does not imply a large increase in *overall* gas utilization; gas plays a larger role in heating than in power generation. Overall CO_2 emissions in the EU rise by 0.7 percent in 2005 and 1.49 percent in 2020, again mainly as a result of increased utilization of coal.

With respect to German GDP, the nuclear phase-out leads to a reduction by somewhat less than 0.6 percent in 2005 and about 0.3 percent in 2020. In the intermediate term, the German current account drops quite significantly, due to increased electricity imports. In the long term, this effect is reversed, as German nuclear power is predominantly replaced by increased domestic generation, especially based on domestic lignite, rather than imports.

Outside of Germany, the macroeconomic effects are in general rather negligible. Only in Belgium/Netherlands is a visible increase in GDP observed, which tends to fade over the course of time. And, in total, the effect on employment is very small, both in Germany and in the rest of the EU.

Table 6.12. Effects of German Nuclear Phase-Out on CO_2 and the Macroeconomy (Percent)

2005	GE	FR	UK	IT	BN	IB	EU
CO_2	2.31	1.54	0.06	-2.18	1.67	-0.25	0.71
Hard Coal	7.74	1.98	0.22	0.20	1.18	-1.13	2.73
Lignite	-0.49	0.00	0.00	0.00	0.00	-0.06	-0.44
Oil	1.74	0.79	-0.08	-1.09	0.20	-0.03	0.37
Gas	-1.38	3.18	0.11	-7.53	2.96	2.14	-0.18
GDP	-0.55	0.01	0.08	-0.28	0.24	-0.14	-0.17
Consumption	-0.48	-0.50	-0.28	-0.30	1.07	-0.57	-0.29
Investment	-0.14	-0.83	-0.35	0.42	1.34	-0.30	-0.15
Export	0.21	0.19	-0.03	-0.35	0.04	0.09	0.04
Import	0.65	-0.25	-0.20	-0.08	-0.01	-0.01	0.06
Employment	-0.01	-0.08	0.07	0.04	0.06	0.04	0.01
Capital Stock	0.07	0.00	-0.01	-0.01	-0.01	0.03	0.02
Exchange Rate	0.03	0.03	0.03	0.03	0.03	0.03	0.03

Table 6.12. (continued)

2020	GE	FR	UK	IT	BN	IB	EU
CO_2	3.46	0.31	0.19	1.38	1.59	0.41	1.39
Hard Coal	6.60	0.70	-0.04	6.32	0.02	1.48	2.93
Lignite	10.87	0.00	0.00	0.00	0.00	-0.11	9.45
Oil	0.85	0.05	-0.11	-0.08	0.13	-0.16	0.17
Gas	1.09	0.67	0.49	0.47	2.86	1.28	1.09
GDP	-0.32	0.01	0.03	-0.08	0.16	-0.13	-0.09
Consumption	-0.35	-0.06	-0.02	-0.26	0.82	-0.81	-0.14
Investment	-0.88	-0.09	0.09	-0.55	0.14	-0.46	-0.35
Export	0.29	-0.18	-0.24	-0.07	-0.01	0.21	0.01
Import	0.20	-0.08	-0.09	-0.30	-0.14	-0.15	-0.06
Employment	0.04	-0.03	-0.03	-0.05	0.08	-0.02	0.00
Capital Stock	-0.10	0.03	0.05	-0.09	-0.10	0.03	-0.03
Exchange Rate	-0.02	-0.02	-0.02	-0.02	-0.02	-0.02	-0.02

6.5 Conclusions

This chapter has considered the effects of a European Internal Market for electricity and a German nuclear phase-out within the context of this Internal Market. In running the simulations it was assumed that competitive pressure arising in the Internal Market leads to a gradual convergence of capital and fuel costs, which currently differ across regions. In addition, it was assumed that the discount rate utilized in the investment calculus is higher in the open market case than in the closed market case, reflecting a premium for increased market risk. Electricity demand is assumed to be unaffected by market opening and national energy policy.

Due to the latter assumption, the opening of national electricity markets results in a relocation of power generation to the various regions and to a change in the fuel mix, but not to a change in overall power generation (net of transmission losses).[6]

As a result of market opening, power generation is relocated from Germany, Italy and Spain/Portugal to France, the U.K. and Belgium/Netherlands. In the long term (2020), power generation may decrease by up to 5 percent in Germany and rise by somewhat more than 11 percent in Belgium/Netherlands. The import shares arising from a scenario of unrestricted electricity trade stay well below the limits implied by the European Commission's Internal Market directive. They do not necessarily increase over time, because competitive pressure present in the

[6] *Gross* power generation may rise slightly with increased electricity trade, due to transmission losses.

Internal Market leads to a partial convergence of costs reducing the incentives for trade.

The effect on electricity-related CO_2 emissions is different in the intermediate term and in the long term. In the intermediate term (2005), there is a relocation of emissions across the various regions in line with the relocation of power generation. Overall, electricity-related emissions increase by almost 3 percent. In the long term (2020), a substantial change in the electricity industry's fuel mix takes place. A strong substitution of gas for coal occurs. As a result, CO_2 decreases not only in those countries, in which electricity generation decreases, but in the others as well. Overall, electricity-related CO_2 drops by almost 7 percent.

With respect to the general economy, the opening of electricity markets leads to changes in GDP, which mirror the changes in power generation: there are increases in France, the U.K. and Belgium/Netherlands and reductions in Germany, Italy and Spain/Portugal. However, the size of the changes is rather modest. In general, the long-term changes are below 0.2 percent.

With respect to CO_2 intermediate-term increases in all regions are recorded. Totalling 2.5 percent for the EU. In the long term, emissions rise only in France and the U.K., and overall European emissions drop by 0.2 percent.

In the second scenario, in which Germany dismantles its nuclear capacities, the resulting gap is partly filled by imports and partly by increasing fossil-based domestic power generation. As a (direct or indirect) result of larger German imports, power generation in all European regions increases. The increase in power generation is mainly based on gas. In Germany, gas-based power generation increases by almost 250 percent in the intermediate term. In addition, there is an increase in coal-based electricity by more than 10 percent. Electricity-related CO_2 increases by 11 percent in Germany and by almost 8 percent in the European average. In the long term, the substitution of imports for German power generation is partly reversed. An increase in lignite-based electricity takes place in Germany, and electricity-related German CO_2 rises by 19 percent.

With respect to other industries, the nuclear phase-out sparks an increase in the German oil, gas and equipment goods industries. In the long term an increase in the lignite industry is also observed. Most other sectors experience a loss. German GDP decreases by about 0.6 percent in the intermediate term and 0.3 percent in the long term. Total German CO_2 emissions increase by 2.3 percent in the intermediate term and 3.5 percent in the long term. Total European CO_2 emissions rise by 0.7 and 1.4 percent in the intermediate and long term, respectively.

Overall, it can be concluded that both the opening of European electricity markets and a hypothetical German nuclear phase-out will have visible effects on European CO_2 emissions. While the nuclear phase-out will lead to more CO_2, as one would have expected, market opening will trigger a long-term substitution of gas for the more carbon-intensive fuels, such that CO_2 may be expected to be reduced, albeit by a small amount.

7 Effects of Carbon/Energy Taxes

7.1 Introduction

In 1992 the European Commission submitted a draft directive suggesting the introduction of a combined carbon/energy tax in the European Union. The tax was meant as a contribution to achieving the EU's target of stabilizing its CO_2 emissions at the level of the year 1990.

The tax contains a CO_2 and an energy component. The tax base for the CO_2 component is the amount of CO_2 released from the burning of fossil energy carriers. The energy component is generally linked to the heat content of the primary energy carriers. However, the electricity industry is treated differently from the other users of primary energy. Whereas the CO_2 tax is linked to the carbon content of the fuels (just as in the other industries), the tax base for the energy tax is electricity produced. Hydro power is exempted from the tax.

The level of the tax was fixed by the provision that one barrel of crude oil should initially be charged at three USD. This initial rate should subsequently rise by one USD annually for six years. Of these amounts, 50 percent should be related to the carbon content and 50 percent to the heat content of crude oil. This gives 2.81 ECU/tCO_2 and 0.21 ECU/GJ as the specific carbon and energy tax rates, to be applied to the various fuels. The initial energy tax on electricity is 2.1 ECU/MWh.

In what follows we adopt this taxation scheme and augment it by the assumption that after the first seven years the basic tax level increases further, by ½ USD per year. To obtain *real* rates of increase, we deflate the nominal rates by 2.5 percent annually.

With respect to the revenue from the combined carbon/energy tax the directive postulates a revenue-neutral, compensating reduction of existing other taxes. In what follows, we consider two variants of revenue-neutral tax recycling. First, we consider a reduction in income taxes, which we model as a lump-sum transfer to private households. A second variant of revenue recycling is the use of tax revenue to lower labor costs. If this is achieved by substituting the revenue from the environmental tax for social security contributions borne by employers, it also conforms to the principle of revenue-neutrality. Technically, we model this as a subsidization of wages. Both methods of revenue-recycling are performed on a country-by-country basis.

It is known that the initiative to introduce a carbon/energy tax on an EU-wide scale recently failed. Instead, the member states are left with the task of introducing such a tax (or a similar one) unilaterally. In this chapter we analyze both the case of a joint, EU-wide introduction of the tax and of unilateral introduction in Germany.

7.2 European Carbon/Energy Tax

The effectiveness of the carbon/energy tax in terms of CO_2 abatement depends crucially on the induced change in energy prices. The induced price change differs across the various energy carriers. The relative price change depends on the absolute tax burden, as implied by the specific carbon content, and on the relative prices before tax. The carbon content is highest with brown coal, followed by hard coal, petroleum, and gas.[1] The price before tax is highest for petroleum (including products such as gasoline), followed by gas, hard coal and brown coal.[2] The high carbon content along with the low price before tax implies that the price increase is largest for brown coal, followed by hard coal, gas and petroleum. For electricity, the carbon/energy tax implies an increase in generation costs by 16 to 19 percent in 2005 and by 33 to 42 percent in 2020.

The induced price changes of the various energy carriers influence the overall (i.e., average) energy prices of the various energy using sectors, depending on each sector's particular energy mix. The effect of the carbon/energy tax on the average energy input prices of the various non-energy sectors is presented in *Table 7.1*. It can be seen that they are rather different across sectors and across regions. Comparing regions, one finds that the percentage energy price increases are relatively high in Germany, Belgium/Netherlands and Spain/Portugal. These are the countries, in which the energy mix is generally quite carbon intensive. This is mainly due to the high share of coal in power generation. In Italy and France, coal plays a minor role. In addition, France has a high share of nuclear energy. These structural features explain, why the price increases are relatively low in Italy and France.

Comparing the various sectors, it can be seen that in Germany, Belgium/Netherlands and Spain/Portugal the energy price increase remains relatively small in agriculture, construction and transportation compared to the other sectors. In France, the U.K. and Italy the various sectors do not differ to such a degree. In these countries the basic materials industry stands out for its relatively strong rise in energy price.

Having described the energy price effects, we now consider the effect of the European carbon/energy tax on the various sectors and on the macroeconomy.

---

[1] The carbon content is as follows: hard coal: 1.06 tC/TOE, brown coal: 1.28 tC/TOE, petroleum: 0.89 tC/TOE, gas: 0.63 tC/TOE (TOE= ton of oil equivalent).

[2] Lignite (domestic) and hard coal (imported) are priced equally.

Table 7.1. Effect of the European Carbon/Energy Tax on Sectoral Energy Prices (Percent)

2005	GE	FR	UK	IT	BN	IB
Agriculture	9.70	3.90	6.09	3.19	12.57	7.45
Basic Materials	11.75	6.02	8.48	12.72	15.36	14.13
Equipment	12.27	4.38	7.83	6.07	16.53	14.67
Consumption G.	11.92	4.45	7.60	6.27	16.82	12.95
Construction	6.27	4.19	8.23	3.89	9.64	8.09
Transport	7.52	4.12	5.98	4.08	7.19	6.06
Private Services	10.95	3.62	6.35	3.69	13.62	12.95
Public Services	11.71	4.56	7.77	4.53	14.29	12.89

2020	GE	FR	UK	IT	BN	IB
Agriculture	16.66	7.74	13.27	15.10	27.48	16.95
Basic Materials	23.78	12.31	18.76	23.06	33.31	24.46
Equipment	22.08	8.50	14.80	21.09	37.69	30.71
Consumption G.	22.05	8.68	14.43	20.97	38.31	26.18
Construction	12.21	8.33	17.58	12.57	20.66	17.37
Transport	13.74	8.33	14.63	9.77	15.06	12.74
Private Services	18.53	6.98	12.63	14.47	30.56	23.83
Public Services	21.27	8.96	16.71	16.17	31.93	25.22

The effect on the electricity industry is displayed in *Table 7.2*. Note that overall European power generation (net of transmission losses) is assumed to be

Table 7.2. Effect of the European Carbon/Energy Tax on Power Generation, Trade and Electricity Related CO_2 Emissions (Percent)

2005	GE	FR	UK	IT	BN	IB	EU
Total Gen.	-7.36	0.41	-1.59	9.60	14.41	-2.47	-0.03
HC	-26.66	-10.05	1.31	-10.10	-8.30	-14.31	-12.74
BC	-0.16	0.00	0.00	-22.56	0.00	0.00	-0.17
Oil	-64.21	1717.18	0.00	-3.77	0.00	0.00	3.94
GAS	11.70	69.82	-5.00	209.24	58.59	16.53	22.75
NUC	0.00	0.00	0.00	0.00	0.00	0.00	0.00
Others	0.00	2.49	0.00	1.02	0.00	4.21	0.86
Imports	64.30	-31.29	0.00	-24.24	-41.83	47.55	-8.07
Exports	-38.73	-2.63	-30.24	0.00	199.30	-48.98	-8.27
CO_2.	-11.45	-1.62	-0.99	0.73	7.74	-8.40	-4.70

Table 7.2. (continued)

2020	GE	FR	UK	IT	BN	IB	EU
Total Gen.	0.60	2.30	2.02	-8.04	8.50	12.06	0.03
HC	-38.82	-33.44	-13.59	-40.58	-42.63	-42.01	-31.12
BC	-11.22	0.00	0.00	-31.50	0.00	-38.19	-14.92
Oil	-97.15	-89.72	0.00	-57.60	-66.39	0.00	-33.79
GAS	90.76	478.01	25.24	29.54	33.62	-4.54	42.47
NUC	0.00	0.00	0.00	0.00	0.00	0.00	0.00
Others	0.00	3.29	0.00	3.29	0.00	-2.63	1.13
Imports	-16.03	104.09	-55.74	54.53	-58.45	384.54	18.10
Exports	-25.64	21.82	143.60	-97.47	33.97	-44.26	18.44
CO_2	-12.82	-14.88	-4.68	-22.24	-5.30	-23.81	-12.78

unaffected by the tax. As a result of this, the impact of the tax is a purely structural one consisting of a relocation of power generation across regions and a change in the energy mix. As can be seen, these two effects are sufficient to stimulate a reduction in European electricity-related CO_2 by almost 5 percent in the intermediate term (2005) and by almost 13 percent in the long term (2020).

The effect on electricity-related CO_2 differs from region to region. It also varies from the intermediate term to the long term. In the intermediate term, the effect ranges from a reduction of more than 11 percent in Germany to an increase of almost 8 percent in Belgium/Netherlands. Correspondingly, we find that power generation is greatly reduced in Germany, whereas a strong increase in Belgium/Netherlands occurs. The reason for this lies, of course, in the high coal dependence of the German electricity system and the high share of gas in Belgium/Netherlands. Accordingly, electricity in Belgium/Netherlands increases its cost advantage relative to Germany resulting in a strong increase in exports to Germany. By contrast, France does not increase its exports, because nuclear capacities are restricted. Actually, French power generation and power exports are reduced. This leads to reduced Italian power imports, with the interesting consequence that Italian power generation rises by almost 10 percent. As a further consequence, Italian CO_2 from power generation rises as well. However, the increase in CO_2 is small with additional electricity stemming mainly from gas, whereas coal and oil based generation is reduced.

A similar change in the fuel mix can be observed in all countries, except for the U.K. In the U.K. the gas price before tax is particularly low, such that the tax makes gas *relatively* more expensive than coal. On the average a strong substitution of gas for coal is undertaken throughout the EU.

Thus, it can be stated that in the intermediate term the carbon/energy tax takes hold mainly through a change in the energy mix in favor of gas. It must, however, be noted that this effect is limited by the persistence of existing power generation capacities. This explains in particular, why lignite-based generation in Germany and Spain/Portugal is minimally affected by the tax. In the long term, however, other options open. Given the age composition of existing plants, major capacity adjustments in the European power supply industry are to be expected beyond 2005 with or without a carbon/energy tax. If such a tax is implemented, coal based capacity, to be dismantled based on plant lifetime, will largely be replaced by gas-fired power plants. As a result, the long term effect on electricity-related CO_2 is much more pronounced than the intermediate term effect.

More specifically, it can be seen that by 2020 a reduction in CO_2 in *all* countries is visible. Interestingly, this occurs in spite of an *increase* in power generation in Germany, France, the U.K. and Belgium/Netherlands. Decreases take place only in Italy and Spain/Portugal; these countries substitute electricity imports for domestic generation. The converse is true in Germany: whereas in the short term Germany reacts to the carbon/energy tax by increasing its imports, the long term response involves an expansion of gas-fired capacity and a reduction of imports. As a result French export capacities are no longer absorbed by Germany, and become available to Italy and Spain/Portugal.

Now we turn to the effect of the carbon/energy tax on other sectors of the economy, as summarized in *Table 7.3*.

It is evident that the energy sectors are most strongly affected by the tax. Considering the EU as a whole, the strongest reductions occur in the hard coal and lignite industries, followed by the oil industry. The gas industry is a gainer - on the European average.

With respect to the non-energy sectors, the effects of the carbon/energy tax largely depend on their energy intensity and on the size of the induced energy price increase. As described in subsection 5.2.1, basic materials and transportation are the most energy intensive industries by far. In addition, basic materials experience the strongest energy price increase. Another factor contributing to the sectoral effects is the demand composition. Because the model assumes a relatively weak substitutability between energy and capital and because the tax revenue is redistributed to the private households, more consumption-oriented sectors can be expected to be less strongly affected than investment-oriented ones.

Considering the effects in the intermediate term (2005), with respect to the EU as a whole, the basic materials and the construction sectors are most strongly affected, whereas the equipment and consumption goods industries remain more or less unaffected. Clearly, the situation differs from country to country. It is especially interesting to observe that the construction industry in the Netherlands and in Spain/Portugal expands, in contrast to all other countries. Basic materials experience a relatively strong decrease in all countries except the U.K., where this industry experiences an expansion. The transportation sector is most strongly affected in Belgium/Netherlands and in Spain/Portugal. Interestingly, almost all non-energy sectors in the U.K. (except for Construction and Public Services) show signs of growth.

Table 7.3. Sectoral Effect of the European Carbon/Energy Tax (Percent)

2005	GE	FR	UK	IT	BN	IB	EU
Agriculture	-0.15	-0.11	0.10	-0.12	-0.34	-0.07	-0.11
Hard Coal	0.00	-13.69	0.51	-42.25	-44.19	-8.41	-7.70
Lignite	-1.23	0.00	0.00	0.00	0.00	-0.18	-0.97
Oil	-1.10	-0.50	-0.52	-1.65	-1.55	-1.11	-0.89
Gas	0.28	-0.79	-2.23	11.06	0.15	8.54	0.37
Electricity	-8.47	0.40	-1.59	9.60	14.41	-2.47	-0.71
Basic Materials	-0.29	-0.36	0.62	-0.55	-1.21	-0.11	-0.28
Equipment	0.16	-0.21	0.54	-0.09	-0.58	0.16	0.07
Consumption G.	0.11	-0.08	0.35	-0.14	-0.40	0.08	0.02
Construction	-0.43	-0.43	-1.05	-0.02	0.42	0.32	-0.34
Transport	-0.09	-0.16	0.16	-0.16	-0.29	-0.28	-0.11
Private Services	-0.12	-0.11	0.02	-0.09	0.13	-0.13	-0.07
Public Services	-0.38	-0.27	-0.10	-0.30	-0.33	-0.54	-0.27

2020	GE	FR	UK	IT	BN	IB	EU
Agriculture	-0.47	-0.32	-0.63	0.09	-0.52	0.69	-0.21
Hard Coal	0.00	-28.28	-12.57	-58.21	-60.90	-17.25	-21.23
Lignite	-7.98	0.00	0.00	0.00	0.00	-32.19	-13.67
Oil	-3.84	-2.82	-1.82	-3.17	-2.49	-3.50	-2.56
Gas	8.96	0.84	-3.72	10.61	-0.02	-2.00	1.96
Electricity	0.60	2.30	2.02	-8.04	8.50	-12.06	0.05
Basic Materials	-0.70	-0.78	0.01	-0.56	-2.03	-0.29	-0.66
Equipment	0.06	-0.21	-0.42	0.32	-1.01	0.38	-0.06
Consumption G.	0.07	-0.05	-0.27	0.35	-0.29	1.22	0.13
Construction	-0.77	-0.36	-2.07	0.21	-0.22	2.24	-0.46
Transport	-0.15	-0.12	-0.41	0.08	-0.30	0.06	-0.15
Private Services	-0.42	-0.03	-0.49	-0.01	0.12	1.44	-0.09
Public Services	-0.88	-0.41	-0.56	-0.80	-0.33	-0.38	-0.64

Overall, with very few exceptions, the intermediate-term output reduction in the non-energy sectors is well below one percent.

In the long term (2020) the picture is, in general, rather similar. Here, the average losses are most pronounced in the basic materials sector, whereas the consumption goods industry expands slightly. Again, the effect on the non-energy sectors are in almost all cases below one percent. In Spain/Portugal most non-energy sectors experience a rather significant long-term gain, especially consumption goods, construction and private services.

We now turn our attention to the effect of the European carbon/energy tax on total CO_2 emissions and on macroeconomic indicators, as shown in *Table 7.4*. The overall CO_2 reduction amounts to almost 4 percent in the intermediate term and about 7.6 percent in the long term. Thus, given the development of emissions to be expected in the absence of CO_2 reduction policies, the tax strays far from achieving a stabilization of European emissions. One should also note that the *overall* percentage CO_2 reduction in the EU is smaller than that in the EU's electricity industry.

Considering individual regions, it is seen that, in the intermediate term the reduction of CO_2 is above average in Italy, Germany and Spain/Portugal and below average in France, the U.K. and Belgium/Netherlands. The low effectiveness of the tax in France is understandable on the basis of the low carbon intensity of the French energy mix and the resulting low energy price increase. In the long term, German CO_2 reduction is slightly below the European average. This mirrors the effect in the German electricity industry, where increased electricity import, as an intermediate-term response, is followed by increased gas utilization as a long-term strategy, implying that the long-term reduction in electricity-related CO_2 is not much larger than the intermediate-term reduction.

Considering the macroeconomic effects, European GDP decreases by about 0.2 percent in the intermediate term and somewhat more than 0.3 percent in the long term. In the intermediate term, the GDP reduction is below average in the U.K., Belgium/Netherlands and in France. In fact, in the U.K. there is even a small *increase* in GDP. In the long term, the GDP reduction is below average in France, Belgium/Netherlands and in Spain/Portugal. Spain/Portugal experiences even a significant *increase* in GDP. Both the intermediate-term increase of the GDP in the U.K. and the long-term increase in Spain/Portugal could have been expected from the sectoral results described above.

Considering the structure of GDP, it is found that on the average investment is more negatively affected than consumption. This tendency stems from the assumed weak substitutability of capital and energy and from the assumed tax-recycling regime, namely the transfer of tax revenue to private households. This tendency is, however, not present in all countries at all dates. Especially, Italy, Belgium/Netherlands and Spain/Portugal experience an increase in intermediate-term investment.

With respect to foreign trade, one observes that in many cases imports are more strongly reduced than exports. This is understandable, since energy has a high share in the European countries' imports, so any reduction of energy utilization has a strong impact on imports. Even though the current account does not improve in *all* countries, it does for the EU as a whole. As a result, the exchange rate declines, i.e., the currency of the EU experiences a revaluation.

The effect of the carbon/energy tax on EU-wide employment is negligible. In the intermediate term, there is a small expansion in the U.K. and in Belgium/Netherlands (less than 0.2 percent); the largest reduction occurs in France (again less than 0.2 peercent). In the long term, employment tends to be negatively affected only in the U.K. (less than 0.3 percent).

Table 7.4. Effect of the European Carbon/Energy Tax on CO_2 and the Macroeconomy (Percent)

2005	GE	FR	UK	IT	BN	IB	EU
CO_2	-5.13	-2.26	-2.70	-5.99	-2.56	-4.37	-3.95
Hard Coal	-17.39	-8.62	-1.47	-9.52	-8.10	-12.02	-10.60
Lignite	-0.60	0.00	0.00	-24.65	0.00	-0.17	-0.58
Oil	-1.08	0.09	-2.34	-2.21	-0.91	-1.19	-1.29
Gas	-1.21	-2.89	-3.49	-12.87	-1.02	-0.15	-3.55
GDP	-0.27	-0.17	0.10	-0.31	-0.02	-0.28	-0.17
Consumption	-0.17	-0.28	-0.81	-0.02	0.98	-0.29	-0.21
Investment	-1.05	-1.00	-2.60	0.27	1.20	0.47	-0.89
Export	-0.19	-0.37	0.28	-0.74	-0.39	-0.40	-0.26
Import	-0.23	-0.58	-0.78	-0.07	-0.30	0.08	-0.38
Employment	-0.03	-0.17	0.19	0.02	0.11	-0.05	0.00
Capital Stock	-0.01	-0.01	-0.24	0.05	0.11	0.36	-0.01
Exchange Rate	-0.24	-0.24	-0.24	-0.24	-0.24	-0.24	-0.24

2020	GE	FR	UK	IT	BN	IB	EU
CO_2	-7.23	-5.84	-6.72	-11.23	-7.31	-8.97	-7.64
Hard Coal	-24.66	-25.55	-14.67	-28.31	-28.23	-25.51	-23.35
Lignite	-7.51	0.00	0.00	-31.34	0.00	-26.00	-9.91
Oil	-3.24	-3.42	-6.17	-7.65	-2.20	-1.62	-4.35
Gas	0.22	3.30	-4.00	-4.64	-3.65	-1.84	-2.33
GDP	-0.74	-0.13	-0.42	-0.46	-0.32	1.02	-0.34
Consumption	-0.75	-0.31	-1.32	-0.74	0.90	3.44	-0.34
Investment	-2.19	-1.37	-4.23	-0.23	-0.86	2.81	-1.72
Export	-0.57	-0.73	-0.35	-0.81	-0.67	-2.10	-0.70
Import	-0.80	-0.95	-1.72	0.17	-0.85	0.29	-0.82
Employment	0.00	0.06	-0.28	0.04	0.09	-0.01	-0.03
Capital Stock	-0.38	-0.26	-1.22	0.31	0.20	1.23	-0.25
Exchange Rate	-0.91	-0.91	-0.91	-0.91	-0.91	-0.91	-0.91

7.3 National Carbon/Energy Tax in Germany

This section considers the case of a carbon/energy tax designed in the same way as above and introduced unilaterally in Germany.

Given the result that a Europe-wide carbon/energy tax triggers a substantial relocation of power generation across countries, at least in the intermediate term, it can be expected that such a result is even more pronounced, if the tax is introduced unilaterally in a single country such as Germany.

Such an effect does indeed arise, and it turns out to be rather substantial, as can be seen from *Table 7.5*. Power generation in Germany is reduced by more than 28 percent in the intermediate term and by almost 58 percent in the long term, and imports increase by several hundred percent. This is accompanied by a reduction in German electricity-related CO_2 emissions by more than 45 percent in the intermediate term and by more than 88 percent in the long term. That CO_2 is reduced more than generation is due to a significant change in the fuel mix: though *all* fossil fuels are reduced, the reduction is substantially more pronounced with coal and oil than with gas. This is true especially in the short term.

Table 7.5. Effect of German Carbon/Energy Tax on Power Generation, Trade and Electricity-Related CO_2 Emissions (Percent)

2005	GE	FR	UK	IT	BN	IB	EU
Total Gen.	-28.51	4.12	0.01	23.87	32.82	2.11	0.08
HC	-63.69	22.59	7.43	2.25	4.78	1.95	-10.99
BC	-29.16	0.00	0.00	0.00	0.00	0.00	-26.21
Oil	-64.21	2048.72	0.00	34.57	0.00	0.00	25.83
GAS	-20.88	73.68	-5.00	180.39	104.05	12.41	26.15
NUC	0.00	0.00	0.00	0.00	0.00	0.00	0.00
Others	0.00	-0.30	0.00	-2.53	0.00	0.12	0.88
Imports	338.84	3.50	0.00	-55.05	6.91	-32.83	29.39
Exports	-51.28	18.68	0.23	7950.88	748.56	173.87	30.15
CO_2	-45.52	34.51	2.56	22.23	29.74	2.56	-6.72

2020	GE	FR	UK	IT	BN	IB	EU
Total Gen.	-57.96	13.19	3.04	18.09	41.64	1.22	0.24
HC	-82.12	0.82	0.00	12.76	119.03	14.37	-3.13
BC	-91.52	0.00	0.00	0.00	0.00	0.00	-80.04
Oil	-97.15	132.41	0.00	18.28	0.00	0.00	8.72
GAS	-86.84	860.71	10.57	29.54	33.62	-4.54	25.80
NUC	0.00	0.00	0.00	0.00	0.00	0.00	0.00
Others	0.00	-0.57	0.00	-2.03	0.00	-0.72	-2.97
Imports	406.54	330.65	-55.74	-73.32	33.64	-20.86	135.32
Exports	-34.75	106.56	245.26	1027.40	310.82	232.38	137.84
CO_2	-88.52	52.05	2.30	18.69	65.23	4.24	-10.68

The increase in German imports affects power generation in all other countries except the U.K. The direct effect consists of a strong increase in

Belgium/Netherlands and in France[3]. Indirectly, Italy and Spain/Portugal increase their power generation as well, since domestic generation becomes more competitive relative to (the now more expensive) imports. These increases are mainly gas-based.

As a result of increased power generation outside Germany, CO_2 emissions rise, a phenomenon usually called the leakage effect of unilateral CO_2 reduction. The increase is substantial in France, Belgium/Netherlands and Italy[4]. In spite of this, EU-wide electricity-related emissions drop by almost 7 percent in the intermediate term and almost 11 percent in the long term.

It is extremely remarkable that, in the intermediate term, overall emissions drop more under unilateral taxation than under joint taxation. This can be explained by the high carbon intensity of German power generation compared to Belgium/Netherlands and France. Because the unilateral carbon/energy tax makes German power generation more uncompetitive than a EU-wide tax does, German coal and lignite based power is, to a much larger extent, replaced by less carbon intensive power imports, leading to a stronger reduction in overall electricity-related CO_2.

The effects of the German carbon/energy tax on the other sectors of the economy are displayed in *Table 7.6.*

Evidently, the fuel industries in Germany experience a strong decline (except for hard coal mining, whose output is fixed exogenously). Outside of Germany, there is a compensating increase of hard coal, oil and gas output.

As may be expected, the effects on the non-energy industries outside of Germany are not very pronounced. Yet, in some cases an expansionary effect is recognizable. For instance, the construction industry in Belgium/Netherlands experiences an intermediate-term increase due to construction activity in the electricity industry.

Visible sectoral effects do, of course, arise in Germany. But they are not necessarily negative, nor are they necessarily worse than in the case of EU-wide taxation. The German equipment and consumption goods industries experience an output expansion both in the intermediate and in the long term, which is more pronounced than under EU-wide taxation.

---

[3] In *percentage* terms, the effect is, of course, smaller in France than in Belgium/Netherlands, because of the higher absolute level of power generation.

[4] The *percentage* increase is particularly strong in France, because of the low *initial* emissions.

Table 7.6. Sectoral Effects of the German Carbon/Energy Tax (Percent)

2005	GE	FR	UK	IT	BN	IB	EU
Agriculture	-0.17	0.04	-0.01	-0.03	-0.43	-0.02	-0.07
Hard Coal	0.00	8.17	5.07	-5.86	-5.39	0.53	0.93
Lignite	-19.39	0.00	0.00	0.00	0.00	0.00	-14.64
Oil	-0.78	0.85	0.06	0.64	-0.17	0.25	0.16
Gas	-0.76	0.00	-0.33	8.91	0.99	6.32	0.78
Electricity	-28.51	4.12	0.01	23.87	32.82	2.11	0.08
Basic Materials	-0.07	0.29	0.11	-0.25	-1.41	0.05	-0.13
Equipment	0.77	-0.11	-0.15	-0.14	-1.55	-0.10	0.13
Consumption G.	0.28	0.01	-0.02	-0.15	-0.84	-0.03	-0.04
Construction	-0.41	-0.05	0.04	-0.06	0.58	-0.04	-0.06
Transportation	-0.11	0.07	0.03	-0.03	-0.59	0.01	-0.07
Private Services	-0.31	0.01	0.02	-0.03	0.39	-0.05	-0.06
Public Services	-0.52	0.06	0.04	0.10	0.52	0.07	-0.09

2020	GE	FR	UK	IT	BN	IB	EU
Agriculture	-0.83	0.07	0.03	-0.30	-0.81	0.16	-0.24
Hard Coal	0.00	2.40	0.54	-4.92	9.60	3.45	0.52
Lignite	-45.77	0.00	0.00	0.00	0.00	0.17	-34.98
Oil	-3.36	0.10	0.12	0.39	-0.56	-1.25	-0.43
Gas	-9.09	4.24	0.30	8.01	0.27	-2.47	-0.82
Electricity	-57.96	13.19	3.04	18.09	41.64	1.22	0.24
Basic Materials	-0.81	0.15	0.13	-0.79	-2.06	0.07	-0.56
Equipment	0.68	-0.61	-0.37	-0.43	-2.87	-0.68	-0.18
Consumption G.	0.25	-0.16	-0.05	-0.30	-1.45	0.30	-0.13
Construction	-1.59	0.17	0.16	0.02	0.08	0.68	-0.23
Transport	-0.47	0.10	0.15	-0.19	-1.08	0.31	-0.18
Private Services	-0.91	0.04	0.08	-0.25	0.02	0.99	-0.20
Public Services	-0.64	0.01	0.11	-0.34	0.41	0.60	-0.16

It is extremely remarkable that, in the intermediate term, overall emissions drop more under unilateral taxation than under joint taxation. This can be explained by the high carbon intensity of German power generation compared to Belgium/Netherlands and France. Because the unilateral carbon/energy tax makes German power generation more uncompetitive than a EU-wide tax does, German coal and lignite based power is, to a much larger extent, replaced by less carbon intensive power imports, leading to a stronger reduction in overall electricity-related CO_2.

Overall, some industries in Germany fare better and others worse under the unilateral than under the joint tax. In contrast to common reasoning, the export-oriented industries tend to be more favorably affected by the unilateral than by the joint tax.

Several factors contribute to this. First, the unilateral tax leaves the rest of the EU's demand for German exports largely unaffected, whereas the joint tax does not. Second, the currency revaluation induced by lower energy imports is weaker in the case of unilateral CO_2 reduction measures than it is under a joint strategy.[5] Therefore, comparative advantage of German industry in the world market is affected less under a unilateral strategy.

A third mechanism limiting negative impacts on German industry is that the cost level of German industry partly depends on the prices of imported intermediate inputs, especially from other countries of the EU. Because the prices of imports from other EU countries are lower, if these countries do not levy the carbon/energy tax, this dampens the negative impact of the unilateral tax on German price competitiveness.[6]

We now consider the effects on the general economy and on the corresponding CO_2 emissions, as shown in *Table 7.7*. In the intermediate term, German CO_2 is reduced by about 13 percent and in the long term by 18 percent. These reductions are about 2 ½ times as strong as under joint taxation. It is clear from the discussion above that this increased effectiveness of the tax (in comparison with joint taxation) stems largely from the electricity sector.

As a result of the relocation of CO_2 intensive activities, emissions increase outside of Germany. This applies especially to Belgium/Netherlands and France and is, again, mainly related to electricity generation. The induced increase in Italian emissions is also quite considerable. In spite of these 'leakage effects' the unilateral German carbon/energy tax achieves a reduction of overall European emissions by about 2 percent.

---

[5] In contrast to the effect on *electricity-related* CO_2, the effect of unilateral taxation on *overall* European emissions is weaker than that of joint taxation, as will be seen below. Therefore, European energy imports are reduced less under unilateral taxation, and the effect on the exchange rate is smaller.

[6] This mechanism would be even more pronounced, if the European countries were assumed to have individual currencies. In such a case, export prices of other European countries would not be affected by revaluation, and German intermediate imports from these countries would be even less expensive. Such a set-up is analyzed in Welsch (1996b). Of course, the second factor mentioned above is no longer effective in such a case.

Table 7.7. Effects of the German Carbon/Energy Tax on CO_2 and the Macroeconomy (Percent)

2005	GE	FR	UK	IT	BN	IB	EU
CO_2	-13.15	3.74	0.56	1.98	4.17	0.53	-2.17
Hard Coal	-37.62	13.59	5.28	2.96	3.23	1.41	-8.08
Lignite	-22.52	0.00	0.00	-0.01	0.00	-0.02	-19.64
Oil	-0.82	1.62	-0.01	4.87	0.50	0.03	1.07
Gas	-0.92	0.46	-0.84	-6.94	7.25	1.64	0.07
GDP	-0.48	0.05	-0.01	-0.05	0.48	-0.06	-0.10
Consumption	-0.85	-0.28	-0.10	-0.08	2.22	-0.31	-0.15
Investment	-1.16	-0.30	-0.07	0.14	2.13	-0.22	-0.23
Export	0.60	0.06	-0.16	-0.29	0.28	-0.04	0.16
Import	0.33	0.03	-0.09	-0.19	0.14	-0.07	0.06
Employment	0.02	0.00	0.03	0.11	0.22	0.05	0.05
Capital Stock	-0.03	0.01	0.01	-0.02	-0.02	0.02	-0.01
Exchange Rate	-0.15	-0.15	-0.15	-0.15	-0.15	-0.15	-0.15

2020	GE	FR	UK	IT	BN	IB	EU
CO_2	-18.01	4.52	0.38	2.74	10.72	1.76	-2.12
Hard Coal	-43.65	0.40	-0.06	10.73	48.79	6.80	-4.26
Lignite	-53.77	0.00	0.00	-0.01	0.00	0.57	-46.74
Oil	-2.59	0.48	-0.13	1.92	0.10	0.41	-0.16
Gas	-2.96	17.44	0.88	-2.32	4.77	-1.69	2.22
GDP	-0.97	0.16	0.08	-0.34	0.08	0.93	-0.20
Consumption	-1.41	-0.35	-0.35	-0.48	2.04	2.76	-0.22
Investment	-3.42	-0.19	-0.24	0.26	1.48	0.47	-0.85
Export	0.86	0.10	-0.46	0.40	0.02	-1.58	0.11
Import	-0.31	0.08	-0.31	0.49	-0.18	-0.71	-0.13
Employment	-0.06	-0.13	0.00	-0.02	-0.07	0.03	-0.05
Capital Stock	-0.69	0.10	0.17	-0.03	-0.16	0.02	-0.14
Exchange Rate	-0.75	-0.75	-0.75	-0.75	-0.75	-0.75	-0.75

Considering the macroeconomy, we find that the adverse effect on German GDP equals about 0.5 percent in the intermediate term and one percent in the long term. Similar to the case of joint carbon/energy taxation (and for the same reasons), investment declines more than consumption. With respect to foreign trade, one now observes an intermediate term *increase* in both exports and imports. The increase is more pronounced for exports than for imports. The increase in imports is the net result of lower fossil-fuel imports and higher electricity imports. In the long term, the reduction of fossil-fuel imports dominates the rise in electricity imports, and overall imports decline. The German current account improves both in the short term and in the long term, and so does the consolidated current account of the EU. Consequently, the EU currency experiences a revaluation, which is smaller than in the case of joint taxationIn general, the other countries of the EU fail to benefit from such a development. The effect on the other countries' GDP is mostly quite small. In the intermediate term, Belgium/Netherlands is a major gainer with respect to GDP. In this region, both consumption and investment increase considerably as a result of the German tax, whereas most other countries experience a small decline of these two indicators. The effect on aggregate exports and imports differs from country to country.

With respect to employment, we find the overall effect to be very small. It tends to be positive in the intermediate term, especially in Belgium/Netherlands. It is worth emphasizing that Germany does *not* experience a reduction in employment. In the long term, a slight tendency towards reduced employment arises. Interestingly, this tendency is *not* more pronounced in Germany than on the average in Europe.

7.4 Alternative Tax Recycling

In the preceding two sections, it has been found that the effect of carbon/energy taxation on employment is rather small (-0.3 to +0.2 percent). These results have been obtained under the assumption that the tax revenue is redistributed to private households in a lump-sum fashion (on a country-by-country basis).

An alternative way of tax recycling, which is currently the subject of intensive public debate, is to use the tax revenue to reduce labor costs. Institutionally, this can be achieved by reducing the social security contributions borne by employers, in which case the measure is a variant of so-called ecological fiscal reform. Technically, the measure is modeled as a subsidization of wages out of the revenue from the carbon/energy tax. This section analyzes the effects of the carbon/energy tax *cum* wage subsidization, assuming the tax is introduced on an EU-wide basis.

It is evident that the effects of such a measure are influenced by the way in which wage claims respond to (induced) changes in employment. Our modeling approach assumes that any increase in employment triggers higher wage claims (see section 2.5). The degree of this response is crucial for the effects of this policy measure; it is captured by the elasticity of annual wage increases with respect to employment (see eq. (2.27)). The default value for this parameter is 1.5, which will be discussed first. For sensitivity, we will also consider the value 0.75.[7]

The natural starting point for discussing the effects of this policy measure is the effect on employment. As it can be seen from *Table 7.8*, there is a clear

[7] Estimates of this parameter and effects of alternative parameter choices are considered in Welsch (1996a). An elasticity of 1.5 is relatively high in view of existing evidence.

increase of employment throughout the EU in each particular region. In the intermediate term, the increase is relatively small in France, which may be explained by the relatively small tax revenue (due to low carbon intensity of energy supply). In the long term, France catches up somewhat, since the tax base and, hence, the revenue available for subsidization of labor increases (see below). In the U.K., by contrast, the long-term employment effect is somewhat smaller than the intermediate-term effect. In general, the intermediate-term effect is in the range of 0.5 to one percent. In the long term, it may reach 1.6 percent.

The effect on GDP is also, in general, positive. The only exception is the intermediate-term effect in Italy, which is negative but close to zero. In most cases, the expansionary effect is more pronounced in the long term than in the intermediate term. Only in Germany and the U.K. it tends to fade over the course of time, the long-term effect in Germany being close to zero.

With respect to the macroeconomic indicators consumption, investment and import, we find that they drop less or increase more than in the case of the lump-sum redistribution of tax revenue to households. The smaller reduction of imports is mainly a consequence of the higher level of economic activity. Since the change of revenue recycling does not affect exports to the same extent as imports, the improvement of the current account is smaller than in the previous case, and the revaluation effect is negligible.

Table 7.8. Effects of the European Carbon/Energy Tax *cum* Wage Subsidization on CO_2 and the Macroeconomy under High Wage Elasticity (Percent)

2005	GE	FR	UK	IT	BN	IB	EU
CO_2	-4.93	-2.02	-2.38	-5.83	-2.34	-4.08	-3.71
Hard Coal	-17.24	-8.45	-1.30	-9.43	-7.94	-11.81	-10.44
Lignite	-0.46	0.00	0.00	-24.65	0.00	-0.02	-0.44
Oil	-0.83	0.34	-2.05	-2.03	-0.67	-0.85	-1.04
Gas	-0.96	-2.57	-3.09	-12.68	-0.79	0.18	-3.24
GDP	0.12	0.20	0.59	-0.03	0.45	0.26	0.23
Consumption	0.20	0.05	-0.50	0.28	1.14	0.23	0.12
Investment	-0.88	-0.82	-2.31	0.43	1.32	0.87	-0.69
Export	-0.07	-0.30	0.35	-0.69	-0.30	-0.24	-0.16
Import	-0.06	-0.41	-0.57	0.06	-0.15	0.35	-0.20
Employment	0.53	0.35	0.83	0.52	0.77	0.85	0.59
Capital Stock	0.07	0.06	-0.13	0.11	0.20	0.49	0.07
Exchange Rate	0.00	0.00	0.00	0.00	0.00	0.00	0.00

Table 7.8. (continued)

2020	GE	FR	UK	IT	BN	IB	EU
CO_2	-6.77	-5.35	-6.10	-10.86	-6.77	-8.24	-7.11
Hard Coal	-24.29	-25.20	-14.40	-28.13	-27.84	-24.87	-23.00
Lignite	-7.18	0.00	0.00	-31.34	0.00	-25.52	-9.56
Oil	-2.73	-2.93	-5.54	-7.22	-1.66	-0.75	-3.79
Gas	0.76	3.92	-3.25	-4.25	-3.05	-1.52	-1.71
GDP	0.03	0.57	0.51	0.13	0.73	2.34	0.48
Consumption	-0.11	0.28	-0.85	-0.21	1.08	4.79	0.25
Investment	-1.68	-0.87	-3.68	0.23	-0.47	3.90	-1.19
Export	-0.19	-0.51	-0.06	-0.57	-0.42	-1.71	-0.41
Import	-0.45	-0.64	-1.34	0.43	-0.55	0.87	-0.48
Employment	0.95	0.90	0.77	0.88	1.35	1.62	0.97
Capital Stock	-0.05	0.04	-0.78	0.56	0.60	1.87	0.11
Exchange Rate	-0.10	-0.10	-0.10	-0.10	-0.10	-0.10	-0.10

A further consequence of the increased activity level is, of course, that emissions are now reduced less. With respect to the EU as a whole, the implied difference in the effectiveness of the carbon/energy tax is about 0.25 percent in the intermediate term and 0.5 percent in the long term. The change in the effectiveness is rather similar across countries.

On a sectoral level, it must first be stated that the change in tax revenue redistribution methods has little effect on the electricity industry. This is because labor costs are rather unimportant in electricity generation, and because their cost share does not vary significantly across technologies. Therefore, changes in the relation of labor costs to other costs has a negligible influence on technology choice and, consequently, on overall input requirements and CO_2 emissions in the electricity supply industry.

Considering the sectoral effects, as displayed in *Table 7.9*, it can be seen that most of them experience an output expansion as a result of the carbon/energy tax *cum* wage subsidization. Exceptions are the construction sector and the basic materials industry in some, but not all, countries. Of course, the fuel industries, except for gas, decline, similar as in the case of tax redistribution to households.

These results have been obtained under a relatively pessimistic assumption with respect to the labor market, i.e., a high elasticity of wage claims with respect to employment. With a weaker response of wage claims the effect on employment and economic activity should be stronger.

Table 7.9. Sectoral Effects of the European Carbon/Energy Tax *cum* Wage Subsidization under High Wage Elasticity (Percent)

2005	GE	FR	UK	IT	BN	IB	EU
Agriculture	0.20	0.13	0.55	0.15	-0.03	0.36	0.22
Hard Coal	0.00	-12.77	1.11	-42.00	-42.95	-7.30	-7.31
Lignite	-1.05	0.00	0.00	0.00	0.00	0.00	-0.80
Oil	-0.87	-0.25	-0.36	-1.46	-1.34	-0.80	-0.68
Gas	0.50	-0.48	-1.76	11.22	0.34	8.72	0.67
Electricity	-7.36	0.41	-1.59	9.60	14.41	-2.47	-0.03
Basic Materials	0.15	0.06	1.12	-0.21	-0.64	0.44	0.17
Equipment	0.59	0.22	1.15	0.19	0.12	0.82	0.54
Consumption G.	0.52	0.29	0.91	0.13	0.09	0.56	0.43
Construction	-0.15	-0.15	-0.67	0.20	0.78	0.82	-0.02
Transport	0.18	0.15	0.52	0.06	0.09	0.14	0.21
Private Services	0.19	0.21	0.44	0.14	0.51	0.32	0.27
Public Services	0.12	0.21	0.47	0.23	0.28	0.32	0.27

2020	GE	FR	UK	IT	BN	IB	EU
Agriculture	0.36	0.26	0.42	0.70	0.36	1.86	0.60
Hard Coal	0.00	-27.30	-11.97	-57.92	-59.32	-15.62	-20.59
Lignite	-7.61	0.00	0.00	0.00	0.00	-32.01	-13.34
Oil	-3.38	-2.37	-1.40	-2.78	-2.15	-2.89	-2.13
Gas	9.42	1.45	-2.91	10.87	0.50	-1.78	2.52
Electricity	0.60	2.30	2.02	-8.04	8.50	-12.06	0.05
Basic Materials	0.24	0.09	1.20	0.23	-0.64	1.01	0.35
Equipment	0.98	0.64	0.88	0.98	0.63	1.79	0.94
Consumption G.	0.86	0.67	0.84	0.93	0.85	2.39	0.98
Construction	-0.18	0.19	-1.36	0.68	0.56	3.43	0.18
Transport	0.38	0.43	0.30	0.52	0.59	0.97	0.47
Private Services	0.21	0.56	0.29	0.47	0.92	2.61	0.58
Public Services	0.01	0.43	0.39	0.10	0.86	1.42	0.32

Table 7.10 shows the effects on CO_2 and on the macroeconomic indicators, if the wage elasticity is 0.75, instead of 1.5. It can be seen that this change in the wage elasticity has a substantial influence on the employment effect. In general, employment now increases by more than one percent. Only in Belgium/Netherlands does the effect remain below one percent, in the intermediate term; it is actually somewhat *smaller* than under the higher elasticity.

The logic of the latter result is related to the changes in economic activity and foreign trade arising in the various countries. It can be seen that under the lower wage elasticity most regions' GDP rises much more than under the high wage elasticity. This is the case especially in Germany and also in the U.K., but not in Belgium/Netherlands. Note also that in the former two countries exports now rise, whereas in Belgium/Netherlands they now experience an even stronger decrease than in the high-elasticity case. The reason for this can be found in the fact that the export-oriented industries in Germany and the U.K. (basic materials, equipment goods, consumption goods) are, on average, more labor intensive than their counterparts in Belgium/Netherlands (see Table 5.1.2).

Table 7.10. Effects of the European Carbon/Energy Tax *cum* Wage Subsidization on CO_2 and the Macroeconomy under Low Wage Elasticity (Percent)

2005	GE	FR	UK	IT	BN	IB	EU
CO_2	-4.65	-1.53	-1.98	-5.45	-2.29	-3.70	-3.38
Hard Coal	-17.04	-8.11	-1.07	-9.22	-7.94	-11.53	-10.23
Lignite	-0.32	0.00	0.00	-24.65	0.00	0.14	-0.29
Oil	-0.51	0.82	-1.65	-1.61	-0.66	-0.41	-0.67
Gas	-0.61	-1.92	-2.63	-12.22	-0.69	0.60	-2.85
GDP	0.61	0.92	1.18	0.64	0.43	0.93	0.80
Consumption	0.58	0.56	-0.30	0.85	0.96	0.86	0.50
Investment	-0.55	-0.45	-1.92	0.93	1.44	1.43	-0.31
Export	0.18	0.02	0.57	-0.40	-0.44	0.00	0.02
Import	0.16	-0.07	-0.34	0.39	-0.30	0.67	0.00
Employment	1.20	1.33	1.51	1.64	0.52	1.90	1.35
Capital Stock	0.23	0.26	0.18	0.29	0.47	0.72	0.29
Exchange Rate	0.50	0.50	0.50	0.50	0.50	0.50	0.50

2020	GE	FR	UK	IT	BN	IB	EU
CO_2	-6.29	-4.78	-5.13	-10.33	-6.51	-7.82	-6.52
Hard Coal	-23.89	-24.79	-13.95	-27.87	-27.75	-24.55	-22.65
Lignite	-6.95	0.00	0.00	-31.34	0.00	-25.39	-9.34
Oil	-2.22	-2.38	-4.54	-6.62	-1.52	-0.23	-3.19
Gas	1.37	4.63	-2.10	-3.65	-2.69	-1.26	-0.95
GDP	0.72	1.32	2.08	0.95	0.94	2.96	1.33
Consumption	0.48	0.89	-0.34	0.43	0.81	5.52	0.77
Investment	-0.98	-0.22	-3.12	0.82	-0.14	4.67	-0.56
Export	0.22	-0.13	0.75	0.01	-0.48	-1.54	-0.01
Import	-0.10	-0.27	-0.73	0.82	-0.64	1.10	-0.16
Employment	1.69	1.72	2.61	2.00	1.20	1.82	1.90
Capital Stock	0.42	0.50	-0.17	0.97	1.20	2.73	0.63
Exchange Rate	0.98	0.98	0.98	0.98	0.98	0.98	0.98

Thus, without strongly responding wage claims, the export industries in Germany and the U.K. benefit more from wage subsidization than those in

Belgium/Netherlands, which implies a change in comparative advantage of these countries.[8]

A consequence of the higher activity level arising in the case of a lower wage elasticity is that the CO_2 reduction is now smaller. If we compare the CO_2 reduction arising in this case with the reduction in the case of a lump-sum transfer of the tax revenue, we find the effect to be about 0.5 percentage points lower in the intermediate term and one percentage point lower in the long term.

The sectoral effects of the carbon/energy tax *cum* wage subsidization under the low wage elasticity are displayed in *Table 7.11*. On the average, all non-energy sectors experience an output expansion. In a country-by-country perspective, Belgium/Netherlands performs relatively poorly, especially with respect to the export industries (see above).

Summarizing this section, it can be stated that a carbon/energy tax, whose revenue is used to lower wage costs, simultaneously achieves a reduction in CO_2 emissions and a significant increase in employment. The effect on employment increases and that on CO_2 decreases as the wage elasticity becomes smaller (with the qualification discussed above).

Table 7.11. Sectoral Effects of the European Carbon/Energy Tax *cum* Wage Subsidization under Low Wage Elasticity (Percent)

2005	GE	FR	UK	IT	BN	IB	EU
Agriculture	0.70	0.59	1.23	0.71	0.29	0.93	0.74
Hard Coal	0.00	-10.59	1.83	-41.44	-43.36	-6.08	-6.83
Lignite	-0.88	0.00	0.00	0.00	0.00	0.00	-0.66
Oil	-0.58	0.22	-0.04	-1.06	-1.38	-0.45	-0.35
Gas	0.81	0.12	-1.25	11.60	0.48	8.90	1.04
Electricity	-7.36	0.41	-1.59	9.60	14.41	-2.47	-0.03
Basic Materials	0.77	0.97	1.91	0.59	-0.45	1.18	0.86
Equipment	1.17	1.17	2.02	0.93	-0.07	1.61	1.23
Consumption G.	1.02	1.04	1.64	0.77	0.07	1.15	1.01
Construction	0.24	0.41	-0.14	0.76	0.76	1.43	0.44
Transport	0.52	0.77	1.00	0.59	0.07	0.63	0.64
Private Services	0.57	0.81	0.93	0.64	0.43	0.87	0.70
Public Services	0.76	1.12	1.02	1.44	0.14	1.45	0.98

[8] It is, of course, far from self-evident that the wage elasticity is the same in all regions. The same objection applies, however, to other model parameters, and it is close to impossible to get reliable parameter estimates on a country-by-country basis.

Table 7.11 (continued)

2020	GE	FR	UK	IT	BN	IB	EU
Agriculture	1.22	0.93	2.05	1.52	1.08	2.68	1.50
Hard Coal	0.00	-26.25	-10.81	-57.50	-59.55	-15.44	-19.99
Lignite	-7.29	0.00	0.00	0.00	0.00	-31.84	-13.06
Oil	-2.95	-1.91	-0.73	-2.25	-2.26	-2.65	-1.61
Gas	9.93	2.12	-1.68	11.25	0.90	-1.59	3.23
Electricity	0.60	2.30	2.02	-8.04	8.50	-12.06	0.05
Basic Materials	1.11	1.10	3.33	1.38	-0.01	1.72	1.45
Equipment	1.72	1.59	3.37	1.99	0.82	2.18	2.00
Consumption G.	1.53	1.46	2.77	1.75	1.10	2.96	1.88
Construction	0.40	0.79	-0.31	1.31	0.66	3.98	0.83
Transport	0.81	1.02	1.61	1.18	0.77	1.26	1.12
Private Services	0.75	1.16	1.58	1.11	0.95	3.20	1.24
Public Services	0.87	1.33	1.94	1.29	0.84	2.03	1.35

7.5 Conclusions

This chapter has analyzed the effect of a carbon/energy tax designed along the lines of the proposal submitted by the European Commission in 1992. Both the case of an EU-wide introduction of such a tax and the unilateral introduction in Germany have been considered.

The first-order effect of the tax is an increase in the prices of the energy carriers and of the average energy prices charged to the various industries. The energy price changes induced by the EU-wide tax do not only show a substantial dispersion across the industries in each particular country, but also across countries, for each particular industry. This indicates substantial differences in the carbon intensities of energy inputs both across sector and country.

In the electricity supply industry, the tax leads to a changing energy mix and a relocation of power generation to the various countries, while the overall level of power generation is kept constant, by assumption. These structural changes lead to a reduction of EU-wide electricity-related CO_2 emissions by almost 5 percent in the intermediate term (2005) and by 13 percent in the long term (2020).

The effects on CO_2 stemming from electricity generation vary across countries and are different in the intermediate and in the long term. In the intermediate term, not all countries reduce their-emissions. This result arises because some countries (those with coal-intensive electricity supply) respond to the tax by increasing their imports, given the inflexibility of existing supply structures. This then promotes higher emissions in those countries increasing their exports.

Typically, these switches go along with an increase in gas-based power generation.

In the long term, restructuring of generation capacities becomes more attractive, and imports lose part of their cost advantage as a response to the tax. Therefore, electricity-related CO_2 is reduced in *all* countries. This is mainly achieved through a further increase in gas firing.

With respect to the non-energy industries, it is found that, on the average, the basic materials and the construction industries experience the strongest output reduction. In the former case, this is due to the high energy intensity, and in the latter case to the dependence on investment demand. The sectoral effects differ from country to country. In most cases, they are negative, but output expansions occur as well. In general, the output effects are below one percent, both in the intermediate and in the long term.

Turning to the general economy, a reduction in EU-wide CO_2 by about 4 percent (2005) and 7.6 percent (2020) is found. This is a significantly smaller relative reduction than in the electricity sector, especially in the long term. Comparing the various countries reveals a considerable dispersion of this effect.

The effect on GDP is a reduction by 0.2 percent in the intermediate term, for the EU as a whole. Across the various countries the effect ranges from a reduction of 0.3 to an increase of 0.1 percent. In the long term, the European average is a reduction by about 0.35 percent, with a range from 0.7 percent reduction to one percent increase. Investment is affected more by the tax than is consumption, and imports decrease more than exports. The latter is a straightforward consequence of the high share of energy in European imports. As a further consequence, there is an improvement of the EU's aggregate current account vis a vis the rest of the world, and a revaluation of the EU's currency.

If the carbon/energy tax is introduced unilaterally in Germany, the relocation of power generation is, of course, even more pronounced. In this case power generation in Germany is reduced by more than 28 percent in the intermediate term and by almost 58 percent in the long term. This goes along with a reduction of German electricity-related CO_2 emissions by more than 45 percent in the intermediate term and by more than 88 percent in the long term. Power generation and related CO_2 emissions outside Germany rise sharply. In spite of this, EU-wide electricity-related emissions drop. In the intermediate term, they drop more under unilateral taxation than under joint taxation. This can be explained by the high carbon intensity of German electricity in comparison with electricity imported from Belgium/Netherlands and France.

The effects of the German carbon/energy tax on the non-energy sectors of the economy are not very pronounced. Visible sectoral effects do, of course, arise in Germany. Overall, some industries in Germany fare better and others worse under the unilateral than under the joint tax. In contrast to common reasoning, the export-oriented industries tend to be more favorably affected by the unilateral than by the joint tax, which is partly related to differences in the effect on the exchange rate.

With respect to the general economy, it is found that German CO_2 is reduced about 2 ½ times more under joint taxation. This increased effectiveness of the tax (in comparison with joint taxation) stems largely from the electricity sector.

As a result of the relocation of CO_2 intensive activities, emissions increase outside of Germany. In spite of these 'leakage effects' the unilateral German carbon/energy tax achieves a reduction of overall European emissions by about 2 percent.

The effect on German GDP is a reduction by about 0.5 percent in the intermediate term and one percent in the long term. The effect on the other countries' GDP is mostly quite small, except for Belgium/Netherlands, which, in the intermediate term, is a major gainer in terms of GDP.

These results are based on the assumption that the revenue from the carbon/energy tax is redistributed to private households. Different results arise, if the revenue is used to lower the costs of labor to employers. It is found that such a measure simultaneously achieves a reduction in CO_2 emissions and a significant increase in employment and GDP. In general, the effect on employment and GDP increases and that on CO_2 decreases as the wage elasticity becomes smaller. Only in the case of Belgium/Netherlands does a lower wage elasticity lead to a weaker increase in employment. This result is related to a change in comparative advantage; because the export-oriented industries in Belgium/Netherlands are less labor intensive than their European competitors (especially Germany and the U.K.), the former benefit less from the reduction of labor costs than the latter. As a consequence of this, exports from Belgium/Netherlands drop, whereas those from Germany and the U.K. rise.

A general insight from this chapter is that the effectiveness of the carbon/energy tax with respect to overall CO_2 emissions in each particular region and in the EU as a whole is largely determined by the structure of the various electricity supply systems. In countries in which rigidities in the structure of the electricity system prevent a short-term change in the fuel mix, the carbon/energy tax induces increasing electricity imports. This is possible, because (within each load block) electricity is a homogeneous good. Therefore, trade is restricted only by transmission costs (transmission losses). It is partly for this reason that the European electricity supply system turns out to be more flexible with respect to CO_2 abatement than the rest of the economy. This manifests itself through larger CO_2 reduction rates in the electricity industry than in the general economy. In addition, it should be noted that these reductions are achieved through structural changes alone, not through changes in overall power generation. If these were taken into account, the CO_2 reduction would be still larger.

With respect to unilateral CO_2 reduction measures, it is mainly the electricity industry, which contributes to their surprisingly strong effectiveness. Since German electricity generation is much more carbon intensive than electricity generation in countries to which power generation would be shifted in the case of a unilateral German carbon/energy tax, the increase in CO_2 outside Germany is much less than the decrease within Germany. As a consequence, the 'leakage effect' is not too pronounced even in spite of a strong substitution of imports for domestic power generation, and the effect of the unilateral tax on overall European CO_2 emissions is surprisingly strong.

These results indicate that a detailed modeling of the electricity supply system is of considerable importance for a careful analysis of CO_2 abatement policies in the context of the Internal Energy Market.

8 Tradeable CO_2 Permits

8.1 Introduction

Complementary to the preceding chapter, where we analyzed the implications of carbon/energy taxes on economic performance and CO_2 emissions, we now consider tradeable CO_2 permits as the key instrument of abatement policy. The main advantage of permits as compared to taxes is that they can guarantee the achievement of exogenous abatement targets simply by restricting the amount of issued emission rights to the desired overall emission level.

In our simulations, we distinguish two scenarios, which reflect different assumptions regarding the scope for coordinated policies within the EU. The first scenario, *EU_1,* refers to a situation of non-coordinated domestic environmental policies, where each of the EU member countries reduces CO_2 over time by a uniform rate. In the second scenario *EU_2* EU countries coordinate their domestic abatement policies in order to achieve the same CO_2 reduction as imposed in scenario *EU_1*.

Standard economic theory suggests that there are overall efficiency gains from coordination, because countries can take advantage of cross-country differences in the domestic marginal costs of abatement. Though we know this qualitative result a priori, we do not know the order of magnitude in an empirical setting. The latter might be crucial for the future direction of EU environmental policy with respect to carbon abatement. A move towards coordination does not necessarily imply that all participants are better off. To make coordination desirable for all member countries (i.e. every individual country should be at least as well off as before coordination) there will be the need for negotiating and implementing a compensation mechanism between gainers and losers. If total gains from coordination are not substantial, these gains might not even pay for the transaction costs associated with the negotiation and implementation of a compensation mechanism. In addition, there might be significant administrative costs related to the establishment of an EU-wide market for emission rights. Given this policy background, one motivation of the following analysis is to investigate the quantitative scope for gains from coordination in carbon abatement.

In the simulations, we assume that the representative households in EU countries are initially endowed with emission rights equal to the emission level of their economic activities in the base-year (in our case: 1985). For the benchmark equilibrium, this assumption implies that emissions are not a scarce good and the corresponding price of emission permits is zero. In the counterfactuals, CO_2 reduction targets are achieved through uniform reduction in emission rights across EU member countries. Coordinated versus non-coordinated environmental policy is captured through the regional scope of emission trading:

EU_1: *Non-coordination - auctioned country-specific permits:* Emission rights are solely tradeable within one country, there is no trade of permits across countries. Hence, each member country has to achieve a uniform percentage reduction at the level of the overall EU reduction target.

EU_2 *Coordination - auctioned permits tradeable across EU countries:* Emission rights can be sold and bought across national borders.

In both scenarios, income from emission sales accrue to the representative household in the respective region. In our numerical simulations, we compute counterfactual equilibria for EU-wide CO_2 reductions of 10 %, 20 % and 30 % as compared to the base-year emission level. In every counterfactual, the implicit carbon tax rate sufficient to meet the exogenous reduction target is provided by the shadow price of the emission constraint, i.e. the permit price.[1] For convenience, we shall refer to the price of carbon permits as a tax throughout this chapter. The use of equal reduction targets across scenarios provides a meaningful basis for welfare analysis. Following the traditional design of policy evaluation in applied general equilibrium analysis, the scenario evaluation is based on the comparison of alternative equilibria: The impacts of CO_2 emission constraints are reported with reference to the benchmark equilibrium, where CO_2 emission constraints are not binding. Thus, all departures from the initial equilibrium can be attributed to the CO_2 reduction policy. In the results below all changes are reported as percentage changes with respect to the benchmark values.

8.2 Basic Considerations for Welfare Analysis

Apart from induced structural change at the sectoral level, such as changes in production, employment, imports or exports, our main focus is on the comparison of welfare effects. We report welfare changes as Hicksian equivalent variation (HEV) of benchmark full income, which is spent on current consumption and savings (future consumption). The equivalent variation is defined as the amount of money that must be given to or taken from the representative consumer in

[1] Permits initially held by the representative consumer are equivalent to a tax, the revenue of which is returned lump-sum to the representative consumer.

order to reach the same utility level as after the change in equilibrium prices.[2] Hence, a positive HEV indicates a welfare improvement, a negative one a welfare loss. In order to provide a meaningful welfare analysis, it is necessary to fix government expenditure.[3]

Welfare effects of carbon taxes can be explained as the composition of substitution and income effects. Typically, these effects arise within a country, as well as between countries. For the sake of simplicity, let us first consider the case of a closed economy. Abstracting from environmental benefits, carbon taxes distort marginal consumption and production decisions in favour of less carbon-intensive goods. As a consequence of this domestic substitution effect, the representative consumer leaves his initial Pareto-Optimum and moves to a lower indifference curve. With tax revenues returned lump-sum to the representative consumer the welfare loss is solely due to the domestic substitution effect. When we consider the case of an open economy, we have to account for additional international substitution effects as well as income effects due to tax exports. For a large open economy, the level of the country's trade affects international prices. Hence, carbon taxes change the terms of trade and induce international substitution effects. The welfare implications of domestic and international substitution effects may have opposite signs. For example, import tariffs of a large fossil fuel importer distort domestic production and consumption decisions (i.e. welfare decreasing domestic substitution effect), but may drive down the world price of fossil fuels, hereby improving the country's terms of trade (i.e. welfare improving international substitution effect).

Income effects arise, if domestic taxes are shifted forward not only to the domestic household, but also to foreign households. For example, a unilateral tax on carbon inputs into the production of traded goods will be not only paid by domestic consumers, but also by foreign consumers. A positive tax export effect constitutes an effective transfer from abroad providing an option for shifting the tax incidence. Obviously, all tax export effects have to sum up to zero at the global level.

Based on Fehr, Rosenberg and Wiegard (1994) the following section visualises the basic considerations involved in the decomposition of welfare effects. *Figure 8.1.* illustrates home country *h* as a producer and exporter of some commodity.

---

[2] The equivalent variation indicates how much money a particular change between equilibria is equivalent to. See Shoven and Whalley (1992) for a discussion of different welfare measures.

[3] In the current model formulation private households finance government expenditure through a lump-sum transfer (see equations 4.17 and 4.18). Government expenditure is assumed to be a constant share of full income throughout the simulations. This specification avoids the simplistic fiscal closure having a major impact on the total costs of CO_2 abatement.

Figure 8.1. Decomposition of Welfare Effects

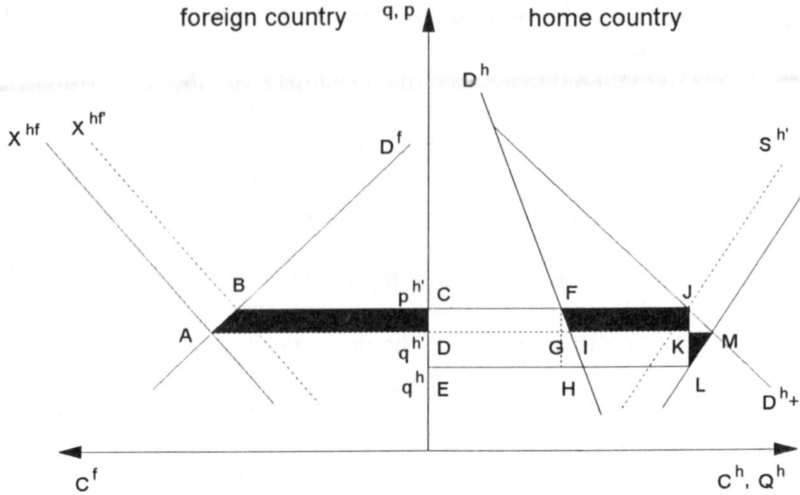

In the right-hand quadrant, S^h is the (inverse) supply curve and D^h is the inverse demand curve of country *h*. The Armington assumption of product heterogeneity implies that these curves do not intersect yielding some excess supply curve which corresponds to the homecountry's export supply curve X^{hf} (see left hand quadrant). In free trade, the equilibrium is determined by the intersection *A* of the export supply and import demand curve (or likewise the intersection *M* of supply S^h and aggregate demand $D^h + D^f$. Taxation of the domestic commodity (i.e. domestic production in the absence of intermediate inputs) implies a leftward shift of the supply curve to $S^{h'}$ and a rightward shift of the export supply curve to $X^{hf'}$ resulting in an after-tax trade equilibrium *B* (or *J* resp.). In the partial equilibrium framework of Figure 8.1., the welfare effects of the tax can be measured through changes in consumer and producer surplus and decomposed into international substitution effect, domestic substitution effects and income (tax export) effect. Starting with the international substitution effect, the tax drives up domestic consumer prices from q^h to $p^{h'}$ and implies a terms-of-trade gain for the home country and a terms-of-trade loss for the foreign country (see area *CBAD*).[4] The domestic substitution effect corresponds to the sum of domestic tax revenues (area *FCEH*) and losses in consumer surplus (area *FCDI*) as well as producer surplus (area *MDEL*). In addition, there is an international income (tax export) effect from taxing the exports to country *f*, which is captured by area *JFHL*. These tax exports constitute an effective transfer of income from the foreign country *f* to home country *h* as tax revenues accrue to the home country *h*, but represent a tax burden to consumers in country *f*. The total

[4] With solely one market and taxation under origin principle, the consumer price can be interpreted as the international terms-of-trade.

domestic welfare change due to the imposition of the tax is given by the difference between area *JFIK* and triangle *MKL*.

For the interpretation of our computational results, it is useful to separate domestic and international substitution effects on the one hand, and income (tax export) effects on the other hand. One way of achieving this is to perform scenario calculations holding relative income levels fixed. With such a partial equilibrium calculation the computational results will report solely the substitution effects of the carbon taxes. The split-down of the aggregate substitution effects into domestic and international substitution components is more sophisticated and omitted in this analysis.[5] Our welfare considerations below are based on changes in HEV (combining substitution and income effects), total substitution effects (combining domestic and international substitution effects) and a composite terms of trade effect. The latter captures international substitution effects as well as income (tax export) effects. We measure percentage changes in the composite ToT on a net-barter basis as follows (see Whalley 1985):

$$\Delta TOT = 100\left(\frac{p_x}{p_m} - 1\right) \tag{8.1}$$

where p_x and p_m are price indices for exports and imports. Here we use a Laspeyres index with the base year's export and import quantities as weights, such that indices for country r' are:

$$p_m = \frac{\sum_{i,r} p^1_{ir}\, M^0_{irr'}}{\sum_{i,r} p^0_{ir}\, M^0_{irr'}} \tag{8.2}$$

$$p_x = \frac{\sum_{i,r} p^1_{ir'}\, M^0_{ir'r}}{\sum_{i,r} p^0_{ir'}\, M^0_{ir'r}} \tag{8.3}$$

where $M_{irr'}$ are country's r' total imports from region r and the additional indices 0 (1) indicate quantities and prices in the benchmark (counterfactual). For normalisation of prices, we use the weighted sum of consumption price indices over regions, where the weights are equal to the region-specific share of

[5] One approach would be to build small open economy models based on the same data and perform separate analysis for each EU country, thereby estimating how much of the welfare effects is due to domestic substitution effects.

benchmark total consumption (i.e. current consumption and savings) in overall EU benchmark total consumption.

8.3 Empirical Results for Non-coordinated Emission Abatement

Table 8.1. presents changes in welfare and macroeconomic indicators as well as the implicit carbon tax rates for different CO_2 reduction targets under a non-coordinated environmental policy.

Table 8.1. Welfare Effects, Carbon Tax Rates and Macroeconomic Effects for Non-Coordinated Abatement

	GE	FR	UK	IT	BN	IB	EU
Welfare Effects (in % HEV from benchmark income)							
10	-0,36	-0,29	-0,39	-0,22	-0,14	-0,18	-0,29
20	-1,12	-0,95	-1,14	-0,81	-0,62	-0,72	-0,95
30	-2,46	-2,17	-2,39	-1,91	-1,6	-1,79	-2,15
Welfare Change due to Substitution (in % HEV from benchmark income)							
10	-0,35	-0,38	-0,20	-0,28	-0,11	-0,30	
20	-1,11	-1,16	-0,75	-0,91	-0,58	-0,94	
30	-2,44	-2,53	-1,78	-2,05	-1,53	-2,07	
Carbon Taxes (in ECU_{85} per ton CO_2)							
10	28,36	38,00	22,94	34,53	20,61	28,75	
20	67,08	91,34	53,34	82,75	49,08	68,26	
30	121,65	168,52	95,35	152,02	89,33	124,44	
Investment (in % change from benchmark)							
10	0,76	0,8	0,39	0,63	0,69	0,74	
20	1,43	1,57	0,64	1,17	1,28	1,36	
30	1,96	2,29	0,69	1,56	1,74	1,8	
Consumption (in % change from benchmark)							
10	-0,79	-0,63	-0,65	-0,55	-0,44	-0,44	
20	-2,09	-1,74	-1,73	-1,55	-1,32	-1,32	
30	-4,12	-3,53	-3,41	-3,2	-2,81	-2,81	
Composite ToT Effect (in % change from benchmark)							
10	0,37	0,48	-0,01	0,45	0,45	0,69	
20	0,77	1	0,02	0,94	0,98	1,47	
30	1,23	1,61	0,1	1,48	1,65	2,41	
ToT for ROW : -0,57 % (10), -1,22 (20), -1,98 (30)							

The negative percentage changes in HEV, which summarize the welfare implications of income and substitution effects, indicate that carbon taxes sufficient to reduce domestic benchmark emissions by 10 %, 20 % or 30 % induce significant welfare losses on the European economy, although all EU member countries can improve their composite ToT (income plus international substitution effect) at the expense of ROW[6]. However, the domestic substitution effect, which dominates the composite ToT effect, makes them worse off as compared to the benchmark situation. Carbon emission constraints reduce the use of fossil fuels in domestic production and consumption and induce substitution towards less carbon-intensive inputs. The domestic substitution effect decreases marginal productivity of factor inputs, which translates into efficiency losses. Efficiency losses through domestic substitution increase towards higher reduction targets as it gets more difficult to substitute for carbon. This is reflected by the magnitude of marginal costs of emission abatement (i.e. the implicit carbon tax rates) ranging between 21 ECU/ton CO_2 (10 % reduction for Belgium/Netherlands) and 169 ECU/ton CO_2 (30 % reduction for France). From Table 8.1., we see that the welfare implications of the overall substitution effect (in our case: mainly a welfare decreasing domestic substitution effect) dominate the welfare implications of the income effects (negative of positive tax exports). Tay exports are negative for the U.K., Belgium/Netherlands and (to a small extent) for Germany, and positive for all other countries. The sign of the income effects is determined by differences in the marginal costs of abatement and the benchmark trade structure. As a rule of thumb, those countries importing large quantities of carbon-intensive goods from countries with higher carbon tax rates should have negative tax exports. For example, Belgium/Netherlands and Germany suffer from tax exports to France, as they import large shares of basic materials from there.

Differences in marginal abatement costs (i.e. carbon tax rates) indicate differences in the "easiness" of CO_2 substitution across regions. The high tax rates for France and Italy stem from the fact that these countries have low elasticities of substitution in sectors with relatively high carbon intensities of the benchmark energy mix (in particular: electricity generation). Hence, higher tax rates are necessary to meet the same exogenous emission reduction target as in other countries (the same argument applies vice versa for countries with low carbon tax rates such as the U. K. and Belgium/Netherlands). Differences in the tax rates across regions indicate a potential for overall efficiency gains from coordination (see section 8.4). Concerning the macroeconomic effects on current

---

[6] Without retaliation of ROW, part of the carbon tax incidence is shifted forward to the EU mainly on the basis of implicit tax exports. Differences in the composite ToT effects can be explained on the basis of the benchmark trade structure between countries and the magnitude of the domestic carbon tax rates. In principle, countries with high tax rates should benefit from tax exports, when they export a lot to countries with relatively lower tax rates. In addition, carbon constraints may improve the ToT for countries which are able to export carbon substitutes.

consumption and investment we see a fall in consumption accompanied by an increase in investment for all EU member countries. These effects stem from the assumption of a constant marginal propensity to save (invest): Due to the composition of current consumption composite carbon taxes increase the consumption cost index more than the cost index for investment. As the investment (i.e. future consumption) becomes cheaper compared to current consumption, the representative consumer substitutes towards the former.

Table 8.2. presents the effects of uncoordinated carbon abatement on sectoral production in different regions. Considering the EU as a whole, production in the fossil fuel sectors is affected most adversely. As mentioned in chapter 7.2, the output effects depend on the change in relative energy prices, which in turn is determined by the specific carbon contents and the initial price differentials per physical energy unit of hard coal, lignite, oil and gas. A high carbon content along with low before-tax prices explains the dramatic cut-back in lignite and hard coal production. Negative output effects for gas and oil indicate substitution away from these fossil fuels in sectoral production and final demand which is not offset by fuel switching from hard coal and lignite to (less carbon-intensive) oil and gas. More specifically, countries with high shares of gas and oil and low shares of coal within the domestic energy mix (see e. g. Belgium/Netherlands) have only limited opportunities to meet the reduction targets by fuel switching. The decrease in overall demand for oil is not significantly higher than that of (less carbon-intensive) gas, because a large part of oil is used in final consumption (private transport and heating) subject to relatively small (own and cross) price elasticities. It should be noted that output effects in regions without significant domestic production of gas and oil represent basically a decline of distribution activities emerging from the reduction of gas and oil imports (see e. g. gas in Italy). Apart from changes in domestic demand, the output effects in the energy sectors of a specific region are strongly influenced by the demand from other regions. This explains the moderate decline in domestic oil production in the U.K., gas production in Belgium/Netherlands and electricity production in France. The U.K. as a net exporter of oil can weaken the negative output effects by shifting oil and gas exports from EU regions to ROW, where no carbon taxes apply. Likewise, Belgium/Netherlands offsets part of the decline in domestic gas demand by increased exports to other EU regions and ROW. As to electricity production, France takes advantage of its carbon extensive electricity generation (nuclear power), compensating the domestic decline with higher exports to other EU countries (see *Tables 8.4. and 8.5.* below).

Table 8.2. Output Effects for Non-Coordinated Abatement (% Change)

		GE	FR	UK	IT	BN	IB	EU
AGR	10	-1,9	-0,6	-1,4	-0,4	-0,6	-1	-1
	20	-4,3	-1,4	-3,2	-1	-1,4	-2,3	-2,2
	30	-7,4	-2,4	-5,5	-1,8	-2,5	-4	-3,7
HC	10	-12	-18	-12	-26	-24	-9,8	-13
	20	-24	-36	-25	-47	-44	-20	-26
	30	-35	-53	-39	-64	-59	-31	-39
BC	10	-19			-22		-9,5	-17
	20	-36			-41		-23	-32
	30	-50			-57		-38	-46
OIL	10	-9	-9,3	-0,2	-7,9	-9,9	-6,8	-5,9
	20	-18	-19	-0,4	-16	-20	-14	-12
	30	-27	-28	-0,7	-24	-31	-21	-18
GAS	10	-3,9	-5,1	-9,2	-1	-1,1	-6,5	-4,6
	20	-9	-12	-18	-2,3	-2,6	-14	-9,9
	30	-16	-20	-27	-4,1	-4,6	-24	-16
EL	10	-3,4	-1,4	-4,8	-4	-3,2	-2,8	-3,3
	20	-7,7	-3,2	-10	-9	-7,1	-6,5	-7,4
	30	-13	-5,6	-17	-15	-12	-11	-12
BAS	10	-4,7	-4,8	-3,9	-3,1	-4,4	-2,8	-4,1
	20	-10	-11	-8,4	-6,8	-9,8	-6,1	-9
	30	-17	-17	-14	-11	-16	-10	-15
EQU	10	0,18	-0,3	-0,4	-0,4	0,02	-0,4	-0,2
	20	0,45	-0,6	-0,9	-1	0,11	-0,8	-0,3
	30	0,84	-0,8	-1,3	-1,6	0,32	-1,2	-0,4
CON	10	-0,5	-0,5	-0,6	-0,5	-0,4	-0,4	-0,5
	20	-1,3	-1,2	-1,3	-1,1	-0,9	-1	-1,2
	30	-2,4	-2,1	-2,3	-2,1	-1,6	-1,9	-2,2
CST	10	0,15	0,38	0,25	0,15	0,21	0,16	0,22
	20	0,15	0,65	0,36	0,15	0,3	0,11	0,3
	30	-0,1	0,79	0,28	-0,1	0,22	-0,2	0,19
TRA	10	-2,7	-3	-4,1	-4,2	-3,6	-5,2	-3,6
	20	-5,9	-6,7	-9	-9,2	-8	-11	-8
	30	-9,8	-11	-15	-15	-13	-18	-13
PRS	10	-0,1		-0,4	-0,1	0,05	0,09	-0,1
	20	-0,3	-0,1	-0,8	-0,3	0,05	0,06	-0,3
	30	-0,7	-0,3	-1,4	-0,7	-0	-0,1	-0,6
PUB	10	0,93	1,01	0,44	0,98	0,95	1,13	0,83
	20	1,83	2,07	0,83	1,97	1,95	2,3	1,64
	30	2,7	3,2	1,15	2,97	2,99	3,51	2,45

As with energy sectors, the output effects on the non-energy sectors are mainly governed by the carbon intensity of the input structure and the initial supply prices of carbon-intensive inputs. Carbon intensive industries such as basic materials or the transport sector have to cope with a sharp increase in input costs, which causes their output prices to increase more and demand for these goods to decrease more as compared to less carbon intensive industries. At the regional level, adverse effects on the output level are enforced for those carbon-intensive industries which are export-intensive and face relatively high carbon tax as compared to other regions (such as e. g. ROW). The severe incidence on the wealth of stock-holders in these industries (due to the sharp decrease in sector-specific capital rents) explains why these industries heavily oppose EU-wide carbon taxes without corresponding action of ROW.[7] As carbon taxes drive down real wages, they change the relative cost structure in favor of less carbon-intensive and more labor-intensive industries. For some industries such as construction or the public service sectors this might even induce an increase of production as compared to the benchmark situation without carbon taxes.

The effects of CO_2 abatement on sectoral employment are presented in *Table 8.3.* Changes in sectoral employment involve the composition of output and substitution effects. The output effect on employment is generally negative (taxes cause a reduction in production and employment). The substitution effect of carbon taxes on employment is generally positive as carbon taxes increase energy prices and reduce the relative wage (marginal productivity of labor). A positive substitution effect lowers the decrease in employment for the case of a negative output effect and strengthens the increase in employment for the case of a positive output effect. The larger the value share of labor, the larger is the substitution effect and the more likely it is that carbon taxes produce an increase rather than a decrease in sectoral employment. Our simulation results show that for some regional industries, the negative output effect can be overcompensated by a positive substitution effect (see e. g. equipment goods in France for the 20 % and 30 % reduction level). Fossil fuel sectors, basic materials and the transport industry face a severe decrease in employment, whereas employment in labor-intensive public services and construction increases.

Changes in imports and exports of EU member countries, which are reported in Tables 8.4. and 8.5., follow the rules of comparative advantage. In a neoclassical model, the location of production is determined by the relative cost of producing goods in a particular country in comparison to the relative cost of producing the same goods elsewhere. The composite effect on imports and exports at the sectoral level of a specific EU country can be split down into trade effects with respect to ROW and trade effects with respect to other EU countries. In general, trade effects with ROW tend to dominate intra-EU trade effects,

---

[7] See Böhringer and Rutherford (1996a, 1996b) or Böhringer, Ferris and Rutherford (1996) for an empirical analysis of issues related to unilateral taxation such as tax exemptions, grandfathered permit systems or voluntary agreements.

because the distortions between world market prices and prices of EU member countries due to EU carbon taxes are stronger than those among EU member countries. As far as trade with ROW is concerned, carbon taxes impose a strong decline in fossil fuel imports due to the decrease of fossil energy utilization in EU countries.[8] Carbon taxes apply to primary energy imports, but they do not apply to embodied carbon in imported goods. With unilateral carbon taxes, those non-fossil fuel sectors in EU countries get less competitive where fossil fuel inputs represent a significant share of direct and indirect costs. This implies that carbon-intensive (non-fossil fuel) goods, such as basic materials, will be imported from the world market rather than produced on the domestic EU markets. Given a strong ROW trade effect, we see a decline of exports in these goods for most countries along with increased imports. Exceptions are basic materials in Belgium/Netherlands, transports in Germany and electricity in France, where intra-EU trade plays a major role. The cost-advantage of nuclear-based electricity in France within the EU increases exports to other EU countries and decreases imports from there.[9] Belgium/Netherlands increases raw material imports from ROW, but this increase is offset by the reduction of imports from the main EU trading partners, which face considerably higher carbon taxes than Belgium/Netherlands. The same applies for transport imports of Germany.

---

8 On the export side the increased gas exports from Germany and France (to ROW, where no carbon tax applies) are due to our Armington trade specification, which allows for crosshauling of data. In absolute figures, these exports are very small and do not matter.

9 Changes in exports and imports of electricity are relatively moderate due to the fact that we use the Armington assumption for traded goods. To represent limited scope for trade in electricity across long distances we use lower elasticities of substitution for trade between EU and ROW as compared to intra-EU trade.

Table 8.3. Employment Effects for Non-Coordinated Abatement (% Change)

		GE	FR	UK	IT	BN	IB
AGR	10	-0,48	-0,05	-0,18	0,11	0,18	-
	20	-1,15	-0,09	-0,42	0,23	0,43	-
	30	-2,1	-0,15	-0,74	0,37	0,79	-
HC	10	-15,2	-20,3	-16,2	-15,6	-21,7	-
	20	-28,8	-37,8	-32,1	-29,3	-39	-
	30	-40,6	-51,8	-46,7	-42,4	-52,7	-
BC	10	-22,4			-26,6		-
	20	-39,5			-47,8		-
	30	-51,7			-62,6		-
OIL	10	-9,06	-9,11	-1,27	-9,17	-8,87	-
	20	-17,9	-18	-2,31	-18,1	-17,7	-16
	30	-26,7	-26,8	-3,07	-26,8	-26,5	-
GAS	10	-5,8	-9,32	-2,38	-2,3	-3,95	-
	20	-12,8	-19,9	-3,67	-5,3	-8,68	-
	30	-21,1	-30,9	-3,9	-9,25	-14,2	-
EL	10	-0,32	-1,42	0;22	0,29	-0,53	-
	20	-0,37	-3,03	0,76	0,79	-0,88	-
	30	-0,06	-4,79	1,73	1,57	-0,94	-
BAS	10	-1,89	-2,08	-1,33	-0,35	-1,84	-
	20	-4,14	-4,52	-2,77	-0,67	-4	-
	30	-6,91	-7,46	-4,41	-0,96	-6,62	-2,1
EQU	10	0,51	-0,05	-0,14	-0,27	0,24	-
	20	1,21	0,03	-0,17	-0,55	0,64	-
	30	2,19	0,29	-0,05	-0,86	1,28	-
CON	10	0,03	-0,05	-0,19	-0,06	-0,01	-
	20	-0,04	-0,14	-0,45	-0,19	-0,07	-
	30	-0,28	-0,31	-0,79	-0,42	-0,19	-0,5
CST	10	0,85	0,99	0,82	0,54	0,71	0,9
	20	1,69	2,02	1,56	0,92	1,38	1,72
	30	2,53	3,11	2,2	1,07	2,01	2,47
TRA	10	-0,93	-0,92	-2,13	-1,32	-2	-
	20	-1,97	-1,91	-4,56	-2,75	-4,27	-
	30	-3,2	-3,01	-7,46	-4,29	-6,88	-
PRS	10	0,09	0,13	0,03	0,25	0,24	0,36
	20		0,16	0,05	0,41	0,45	0,62
	30	-0,33	0,06	0,06	0,45	0,63	0,71
PUB	10	1,33	1,37	0,83	1,36	1,21	1,49
	20	2,76	2,9	1,66	2,85	2,55	3,12
	30	4,33	4,67	2,52	4,52	4,04	4,97

Table 8.4. Import Effects for Non-Coordinated Abatement (% change)

		GE	FR	UK	IT	BN	IB
AGR	10	1,6	0,3	1,5	-0	0,3	2,3
	20	3,2	0,2	3	-1	0,3	4,5
	30	4,9	-1	4,7	-2	0	6,5
HC	10	-21	-22	-23	-18	-12	-21
	20	-36	-38	-41	-32	-22	-36
	30	-47	-49	-55	-43	-31	-48
BC	10	-16			-2	-5	-0
	20	-29			-4	-11	-1
	30	-40			-6	-15	-2
OIL	10	-10	-9	-9	-9	-10	-8
	20	-19	-19	-17	-19	-20	-16
	30	-29	-28	-25	-28	-29	-24
GAS	10	-6	-6		-10	-20	-4
	20	-13	-14		-23	-40	-8
	30	-22	-23		-37	-58	-12
EL	10	1,6	-8	8,2	5,4	2,3	4,5
	20	3,8	-17	20	13	5,8	11
	30	7,2	-28	37	23	11	21
BAS	10	1,1	0,2	0,4	1,2	-1	0,6
	20	2,6	0,7	0,9	2,6	-1	1,3
	30	4,8	1,5	1,8	4,6	-2	2,3
EQU	10	-0	0,5	0,1	0,8	0,2	0,6
	20	-1	0,7	-0	1,4	0,1	0,9
	30	-1	0,7	-1	1,8	-0	0,8
CON	10	-0	-0	-0	0,1	-0	0,5
	20	-1	-1	-1	-0	-1	0,5
	30	-2	-1	-2	-1	-2	-0
CST	10	1,2	-2		-1	0,4	
	20	2,1	-5		-4	0,3	
	30	2,6	-9		-8	-1	
TRA	10	-0	3,2	1,8	5,4	1,5	5
	20	-1	7,3	3,8	13	3,3	11
	30	-1	13	6,2	23	5,8	20
PRS	10	-2	-1	-0	-1	-2	-1
	20	-5	-4	-2	-2	-4	-3
	30	-10	-8	-4	-6	-8	-6
PUB	10	0,9		-0			
	20	1,8		-0			
	30	2,6		-1			

Table 8.5. Export Effects for Non-Coordinated Abatement (% change)

		GE	FR	UK	IT	BN	IB
AGR	10	-3,4	-0,6	-2,6	-0,1	-0,3	-2,5
	20	-7,4	-1,2	-5,6	0,11	-0,7	-5
	30	-12	-1,7	-9,3	0,81	-1,2	-7,9
HC	10	-6,6	-13	-3,8	-35	-30	-13
	20	-14	-29	-7,6	-60	-52	-26
	30	-22	-47	-13	-77	-68	-37
BC	10	0,74					
	20	0,68					
	30	-0,2					
OIL	10	-1,1	-3,4	3,73	1,5	-8,4	0,97
	20	-2,3	-6,7	7,52	3,15	-17	1,97
	30	-3,9	-10	11,4	4,87	-27	2,93
GAS	10	11,4	5,26	-33		9,97	-2,3
	20	26,6	10,5	-60		22,9	-6,1
	30	46,1	13,9	-78		39,1	-13
EL	10	-1,9	4,15	-13	-9,9	-7,8	-6,3
	20	-4,6	9,98	-27	-21	-17	-14
	30	-7,8	18,3	-42	-34	-28	-25
BAS	10	-6,9	-7,2	-6	-6,9	-5	-6,1
	20	-15	-16	-13	-15	-11	-13
	30	-24	-25	-21	-23	-18	-21
EQU	10	0,25	-0,5	-0,2	-0,9	0,07	-0,6
	20	0,81	-0,8	-0,2	-1,8	0,3	-0,9
	30	1,75	-0,9	-0	-2,5	0,76	-0,9
CON	10	-0,8	-0,9	-0,6	-1,1	-0,5	-1,4
	20	-1,7	-1,7	-1,1	-2	-1,1	-2,5
	30	-2,6	-2,4	-1,5	-2,8	-1,8	-3,3
CST	10	-1,8	-2	-1,6	-1,7	-1,6	-3
	20	-3,6	-3,9	-2,6	-3	-3	-5,8
	30	-5,2	-5,7	-3,2	-3,8	-4	-8,5
TRA	10	-5	-8,6	-6,8	-11	-6,1	-13
	20	-11	-18	-14	-23	-13	-26
	30	-17	-29	-23	-37	-21	-40
PRS	10	1,67	1,03	0,36		0,64	0,05
	20	4,72	3,3	1,47	1,14	2,26	1,48
	30	9,8	7,35	3,55	3,87	5,32	4,84
PUB	10	0,02		0,01	0,34	0,17	
	20	0,05		0,05	0,95	0,48	
	30	0,11		0,12	1,92	0,98	

8.4 Empirical Results for Coordinated Emission Abatement

EU-wide cost effectiveness of carbon abatement suggests that marginal costs of emission reduction should be equalized across sources.[10] With respect to overall welfare, a tradeable permit system which leads to a uniform carbon tax across regions should be pareto-superior to a uniform percentage reduction across member countries, where differences in the marginal costs of abatement are present. The welfare numbers in *Table 8.6.* confirm our economic intuition. Under coordinated action EU-wide welfare losses are reduced. However, the scope for improving cost-efficiency of carbon abatement through permit trade seems to be rather limited. For a 10 %, 20 % or 30 % emissions reduction the gains of coordination range well below 5% of the total costs of the uncoordinated case. The point that there are not sufficient differences in the marginal costs of abatement to produce large efficiency gains from permit trading across regions is confirmed through further sensitivity analysis on substitution patterns and elasticity (of substitution) values.[11] At the country level the move towards coordination need not be welfare improving for all EU countries. Aggregate welfare changes are the outcome of income and substitution effects, the welfare implications of which might be opposite. Countries whose marginal abatement costs under uncoordinated policy are below the uniform tax rate will sell permits and abate more emissions (here: the U.K. and Belgium/Netherlands). In turn, countries whose marginal abatement costs under uncoordinated policy are above the uniform tax rate will buy permits and abate less emissions (here: in particular France and Italy).[12] Income and substitution effects are opposite for either group. Permit-sellers face higher carbon taxes after coordination. This worsens the welfare losses due to the domestic substitution effect, but on the other hand improves their composite terms of trade (international substitution effect plus tax export effects). Permit-buyers face lower carbon taxes after coordination. This lowers the welfare losses due to the domestic substitution effect, but worsens the

[10] Our "sub-global" model does not allow for global cost efficiency analysis. We fail to account for potentially important trade-related impacts on emissions outside the EU (e.g. in US, Japan or China). The cross-comparison between uncoordinated and coordinated carbon abatement is predicated on the assumption that leakage effects are of a second-order magnitude.

[11] Due to a lack of country-specific data, nesting structures and elasticities of substitution are assumed to be identical across regions throughout our sensitivity analysis. The one exception is the electricity sector, where elasticities of substitution are implicit and vary across regions due to the bottom-up activity analysis representation of generation technologies.

[12] For Germany and Spain/Portugal, emission trading is less of an issue, since domestic carbon tax rates before coordination are only slightly higher than after coordination.

composite terms of trade effect.[13] In addition, tradeable permits constitute a net income transfer from permit-buyers to permit-sellers, which is not captured by our composite ToT measure.

Table 8.6. Welfare Effects, Carbon Tax Rates and Macroeconomic Effects for Coordinated Abatement

	GE	FR	UK	IT	BN	IB	EU
Welfare Effects (in % HEV from benchmark income)							
10	-0,35	-0,25	-0,4	-0,21	-0,09	-0,18	-0,28
20	-1,12	-0,86	-1,15	-0,78	-0,49	-0,73	-0,92
30	-2,46	-1,99	-2,38	-1,85	-1,35	-1,8	-2,08
Welfare Change due to Substitution (in % HEV from benchmark income)							
10	-0,35	-0,15	-0,39	-0,13	-0,40	-0,30	
20	-1,12	-0,62	-1,20	-0,55	-1,22	-0,94	
30	-2,47	-1,55	-2,61	-1,39	-2,62	-2,06	
Carbon Taxes (in ECU_{85} per ton CO_2)							
10	27.67	27.67	27.67	27.67	27.67	27.67	
20	65,82	65,82	65,82	65,82	65,82	65,82	
30	120,1	120,1	120,1	120,1	120,1	120,1	
Investment (in % change from benchmark)							
10	0,73	0,54	0,54	0,47	1,03	0,71	
20	1,38	0,96	1,02	0,8	2,07	1,28	
30	1,9	1,23	1,42	0,93	3,12	1,67	
Consumption (in % change from benchmark)							
10	-0,77	-0,5	-0,71	-0,47	-0,5	-0,43	
20	-2,07	-1,43	-1,87	-1,37	-1,43	-1,3	
30	-4,1	-2,99	-3,62	-2,89	-2,95	-2,78	
Composite ToT effect (in % from benchmark)							
10	0,35	0,27	0,09	0,29	0,7	0,65	
20	0,72	0,47	0,31	0,53	1,6	1,38	
30	1,14	0,59	0,72	0,71	2,79	2,24	
ToT for ROW : -0,55 % (10), -1,18 (20), -1,93 (30)							

Comparing the aggregate welfare effects under coordination with those under non-coordination we see that coordination is beneficial or equal for most countries except for the U.K. at the 20 % and 30 % reduction level. Within the

[13] As to Germany, for reduction targets of 20 % and 30 % the welfare-improving decrease in the domestic substitution effect is slightly offset by a welfare-decreasing international substitution effect.

group of permit-buyers, Italy and France gain, as the reduced distortions on the domestic markets more than compensate the loss in the ToT and the additional income transfer through permit purchases. Within the group of permit-sellers, Belgium/Netherlands is better off under coordination, whereas for the U.K. coordination at the 20 % and 30 % reduction level imposes additional welfare costs, because the welfare loss due to an increased domestic substitution effect is not compensated through equivalent gains in the ToT and positive transfers through permit sales. Tax export effects - as compared to uncoordinated action - are reversed for Italy and France, which are large buyers of emission permits, and for the U. K., which is a large seller of permits.

Shifts between aggregate consumption and investment across different regions are in first order determined by the changes of carbon tax rates. Higher carbon tax rates typically increase the price index of current consumption as compared to the price index for investment and therefore shift household expenditure towards future consumption.

Output and employment effects under coordinated action (see *Tables 8.7.* and *8.8.*), as compared to uncoordinated action, reflect the changes in the domestic substitution effect. Permit-buyers lower carbon tax rates through coordination, which alleviates the negative output effects in carbon-intensive industries. Labor-intensive industries, such as public services and construction, lose from coordination, because real wage rates are not decreased as much as in the non-coordination case. The opposite picture applies for permit-sellers who face higher domestic substitution effects due to higher carbon tax rates under coordination. With respect to the welfare analysis of coordinated versus uncoordinated action, adjustment costs of labor could provide an additional argument for coordination if these costs were substantially lower under coordination as compared to non-coordination. Our model framework does not account explicitly for labor adjustment costs.[14]

Yet, economy-wide labor displacement in the coordinated case is in the same order of magnitude as in the uncoordinated case, not strengthening the efficiency argument for coordination.[15]

With respect to foreign trade (see *Tables 8.9.* and *8.10.*), countries which buy emission permits have to reduce less emissions as compared to non-coordinated action and, therefore the decline in fossil fuel imports is less distinct. Likewise imports of basic materials or transport services increase less or may even fall (due to increased intra EU-trade), whereas export activities in carbon-intensive goods are affected less adversely. We obtain the reflected image of foreign trade effects for the case of countries selling emission permits.

---

[14] Adjustment costs of physical labor in the short and medium run are roughly captured by making part of the initial capital stock sector-specific.

[15] Countries with higher carbon tax rates after coordination suffer from higher labor friction, whereas countries with lower carbon tax rates reduce their sectoral labor friction as compared to the non-coordinated case.

Table 8.7. Output Effects of Coordinated Abatement (% Change)

		GE	FR	UK	IT	BN	IB	EU
AGR	10	-1,8	-0,3	-1,8	-0,3	-1,1	-1	-0,9
	20	-4,2	-0,7	-4,2	-0,6	-2,4	-	-2,1
	30	-7,2	-1,2	-7,2	-1,1	-4,2	-	-3,7
HC	10	-12	-14	-14	-22	-30	-	-13
	20	-23	-28	-29	-40	-51	-19	-27
	30	-35	-43	-45	-57	-66	-30	-40
BC	10	-19			-18		-	-16
	20	-36			-36		-22	-32
	30	-49			-51		-37	-46
OIL	10	-8,9	-7,1	-0,2	-6,6	-12	-	-5,4
	20	-18	-15	-0,5	-14	-24	-13	-11
	30	-27	-23	-0,8	-21	-36	-20	-17
GAS	10	-3,8	-3,7	-11	-0,8	-1,4	-	-4,8
	20	-8,9	-8,6	-21	-1,9	-3,3	-14	-10
	30	-16	-15	-30	-3,3	-5,6	-23	-16
EL	10	-3,3	-1	-5,7	-3,2	-4,2	-	-3,3
	20	-7,5	-2,3	-13	-7,3	-9,2	-	-7,4
	30	-13	-4,1	-20	-12	-15	-11	-12
BAS	10	-4,5	-3,1	-4,9	-2,3	-6,6	-	-4
	20	-10	-6,9	-11	-5	-14	-	-8,7
	30	-17	-11	-18	-8,4	-23	-	-14
EQU	10	0,1	-0,2	-0,6	-0,3	-0,1	-	-0,2
	20	0,4	-0,2	-1,3	-0,6	-0,2	-	-0,3
	30	0,8	-0	-2,2	-0,9	-0,3	-	-0,3
CON	10	-0,5	-0,3	-0,7	-0,3	-0,6	-	-0,5
	20	-1,3	-0,7	-1,7	-0,8	-1,4	-1	-1,1
	30	-2,4	-1,3	-3,1	-1,4	-2,5	-	-2,1
CST	10	0,1	0,24	0,35	0,11	0,33	0,1	0,22
	20	0,1	0,34	0,63	0,06	0,6	0,0	0,3
	30	-0,1	0,25	0,79	-0,2	0,77	-	0,19
TRA	10	-2,6	-2,1	-5,1	-3,3	-5	-5	-3,6
	20	-5,8	-4,7	-11	-7,3	-11	-11	-7,9
	30	-9,7	-7,9	-19	-12	-17	-18	-13
PRS	10	-0,1	0,02	-0,5	-0	0,01	0,0	-0,1
	20	-0,3	-0	-1	-0,2	-0	0,0	-0,3
	30	-0,7	-0,2	-1,9	-0,4	-0,2	-	-0,6
PUB	10	0,9	0,7	0,57	0,75	1,35	1,0	0,79
	20	1,7	1,38	1,14	1,46	2,83	2,2	1,57
	30	2,6	2,03	1,73	2,12	4,49	3,3	2,36

Table 8.8. Employment Effects of Coordinated Abatement (% Change)

		GE	FR	UK	IT	BN	IB
AGR	10	-0,45	0,07	-0,3	0,16	-0,05	-0,01
	20	-1,09	0,25	-0,76	0,42	-0,16	0,02
	30	-2	0,57	-1,47	0,78	-0,35	0,07
HC	10	-14,8	-15,9	-18,8	-13	-26,2	-12,4
	20	-28,3	-30,9	-36,7	-24,9	-45,2	-24
	30	-40,2	-44,1	-52,3	-36,6	-59,3	-34,6
BC	10	-22			-22,3		-19
	20	-39,1			-42		-39,6
	30	-51,4			-57,3		-57,5
OIL	10	-8,89	-7,04	-1,35	-7,61	-10,4	-7,99
	20	-17,7	-14,4	-2,45	-15,4	-20,3	-15,7
	30	-26,4	-22	-3,27	-23,3	-30	-23,1
GAS	10	-5,7	-6,92	-2,65	-1,88	-4,89	-5,47
	20	-12,7	-15,1	-3,76	-4,28	-10,4	-11,9
	30	-21	-24,5	-3,45	-7,42	-16,5	-19,6
EL	10	-0,3	-0,96	0,28	0,27	-0,74	-0,25
	20	-0,34	-2,1	1,02	0,72	-1,13	-0,32
	30	-0,02	-3,42	2,32	1,4	-1,06	-0,11
BAS	10	-1,8	-1,08	-1,95	-0,01	-3,29	-0,66
	20	-3,98	-2,18	-4,32	0,17	-7,19	-1,26
	30	-6,7	-3,37	-7,21	0,57	-11,9	-1,83
EQU	10	0,49	-0,02	-0,22	-0,18	0,25	-0,23
	20	1,19	0,22	-0,45	-0,26	0,53	-0,39
	30	2,16	0,8	-0,7	-0,25	0,9	-0,5
CON	10	0,03	-0,01	-0,25	-0,03	-0,08	-0,06
	20	-0,06	-0,01	-0,62	-0,06	-0,28	-0,19
	30	-0,3	0,01	-1,16	-0,11	-0,64	-0,45
CST	10	0,83	0,66	1,1	0,37	1,06	0,86
	20	1,65	1,26	2,27	0,55	2,18	1,64
	30	2,48	1,78	3,54	0,45	3,38	2,33
TRA	10	-0,9	-0,56	-2,71	-0,99	-2,81	-1,93
	20	-1,92	-1,1	-5,98	-2,04	-5,98	-4,04
	30	-3,13	-1,61	-10	-3,15	-9,63	-6,36
PRS	10	0,09	0,08	0,03	0,2	0,32	0,35
	20		0,09	0,03	0,33	0,61	0,6
	30	-0,32	-0,02	-0,01	0,38	0,86	0,69
PUB	10	1,3	0,96	1,06	1,06	1,71	1,44
	20	2,7	1,96	2,24	2,17	3,64	3
	30	4,26	3,05	3,59	3,37	5,9	4,77

Table 8.9. Import Effects of Coordinated Abatement (% Change)

		GE	FR	UK	IT	BN	IB
AGR	10	1,52		1,99	-0,2	0,59	2,19
	20	3,07	-0,6	4,32	-1	1	4,22
	30	4,69	-1,9	7,13	-2,6	1,17	6,06
HC	10	-20	-16	-27	-15	-15	-20
	20	-35	-31	-46	-28	-26	-36
	30	-46	-43	-58	-38	-35	-47
BC	10	-16			-1,9	-7	-0,3
	20	-29			-3,8	-13	-0,9
	30	-40			-5,6	-17	-1,6
OIL	10	-9,4	-7,2	-10	-7,8	-12	-7,7
	20	-19	-15	-19	-16	-23	-15
	30	-29	-23	-28	-24	-34	-23
GAS	10	-5,8	-4,8		-8,1	-25	-3,7
	20	-13	-11		-18	-47	-7,7
	30	-21	-18		-31	-64	-12
EL	10	1,47	-8,3	10,4	4,18	3,55	4,05
	20	3,62	-19	26,4	9,74	9,5	10,2
	30	6,84	-32	51,3	17,5	19,3	19,7
BAS	10	0,86	-0,5	0,76	0,55	-0,7	0,5
	20	2,15	-1	1,9	1,19	-1,5	1,1
	30	4,11	-1,4	3,65	2,02	-2,4	1,92
EQU	10	-0,1	0,27	0,18	0,54	0,28	0,6
	20	-0,5	0,2	0,24	0,77	0,46	0,82
	30	-1,4	-0,4	0,18	0,57	0,47	0,57
CON	10	-0,1	-0,2	-0,2	-0	-0,2	0,41
	20	-0,6	-0,7	-0,7	-0,5	-0,6	0,36
	30	-1,5	-1,8	-1,6	-1,5	-1,3	-0,3
CST	10	1,12	-1,6		-1,3	0,92	
	20	1,85	-4,1		-3,6	1,64	
	30	2,13	-7,8		-7,3	2,07	
TRA	10	-0,2	1,55	2,43	4,02	3,15	4,75
	20	-0,5	3,19	5,42	9,22	7,19	10,8
	30	-1	5,09	9,24	16,3	12,7	19
PRS	10	-1,9	-1,3	-0,4	-0,8	-1,8	-0,7
	20	-5,1	-3,8	-1,2	-2,7	-4,4	-2,6
	30	-9,9	-7,7	-2,7	-6,2	-8	-6,1
PUB	10	0,89		0,02			
	20	1,74		-0,2			
	30	2,55		-0,6			

Table 8.10. Export Effects of Coordinated Abatement (% Change)

		GE	FR	UK	IT	BN	IB
AGR	10	-3,3	-0,1	-3,4	0,27	-0,9	-2,3
	20	-7,2	0,11	-7,6	1,04	-2	-4,7
	30	-12	0,64	-13	2,49	-3,3	-7,2
HC	10	-5,8	-14	1,96	-31	-32	-8,2
	20	-13	-28	0,9	-55	-55	-18
	30	-20	-42	-6,2	-72	-72	-29
BC	10	0,48					
	20	0,39					
	30	-0,4					
OIL	10	-1,3	-3,4	4,64	0,91	-9,2	0,76
	20	-2,7	-6,5	9,18	2,3	-19	1,67
	30	-4,4	-9,3	13,7	4,14	-30	2,65
GAS	10	11,8	0,74	-39		15	-2,2
	20	27,9	2,58	-67		33,3	-5,8
	30	48,6	5,26	-84		54	-12
EL	10	-1,2	3,82	-15	-7,7	-9	-6,1
	20	-2,7	9,61	-32	-17	-20	-14
	30	-4,3	18,5	-49	-27	-34	-24
BAS	10	-6,8	-5,1	-7,6	-5,4	-7,2	-5,9
	20	-15	-11	-16	-11	-15	-13
	30	-24	-18	-27	-18	-25	-20
EQU	10	0,24	-0,3	-0,4	-0,6	-0,1	-0,6
	20	0,8	-0,2	-0,7	-1	-0,1	-0,8
	30	1,75	0,48	-1,1	-1	-0,2	-0,8
CON	10	-0,8	-0,6	-0,9	-0,7	-0,9	-1,3
	20	-1,6	-0,8	-1,8	-1,1	-1,9	-2,3
	30	-2,5	-0,6	-3	-1,2	-3,3	-2,9
CST	10	-1,7	-1,3	-2,2	-1	-2,3	-2,7
	20	-3,4	-2,2	-4,3	-1,2	-4,6	-5,2
	30	-4,9	-2,4	-6,4	-0,6	-7	-7,4
TRA	10	-4,9	-6,2	-8,3	-8,8	-8,3	-12
	20	-10	-13	-18	-19	-18	-25
	30	-17	-21	-29	-30	-28	-39
PRS	10	1,64	1,05	0,22	0,34	0,44	0,12
	20	4,68	3,5	0,96	2	1,61	1,68
	30	9,77	8	2,33	5,56	3,85	5,24
PUB	10	0,02		0,01	0,37	0,19	
	20	0,05		0,04	1,03	0,51	
	30	0,11		0,09	2,09	1,02	

8.5 Conclusions

The simulations of EU-wide abatement strategies suggest that carbon abatement imposes significant economic costs independent of coordinated or uncoordinated action. This result is not surprising, because our benchmark economy represents a first-best world without prior distortions. Abstracting from environmental benefits and initial market distortions such as taxes, rigid wages etc., a restriction of carbon or fossil energy necessarily makes the economy worse off if there are no zero-cost substitution options. With unilateral abatement, EU countries can shift part of the carbon tax incidence to the non-abating rest of the world mainly due to tax export effects. Confirming economic intuition, coordination of abatement among EU countries is less costly than uncoordinated action. However, the numerical results indicate that the benefits from coordination are rather low, i.e., marginal costs of abatement across countries are not large enough to yield significant gains from equalization of these costs through permit trade. With winners and losers, coordination is not a no-regret strategy for all countries unless side-payments assure that each country is better off under cooperation than without international cooperation. This implies negotiation on compensating mechanisms, which might involve significant transaction costs. In view of the little overall gains from coordination, advocates of cooperation should be aware that these gains might not pay for transaction costs of negotiation and/or even the implementation costs of a common permit market.

There are a number of important issues related to the analysis of carbon abatement (or more general: environmental regulation) which have not been addressed in our simulations. First, for the sake of modeling simplicity we have ignored public finance issues. As is well-known in economic literature (see e. g. Goulder 1994) the gross costs of environmental taxes depend crucially on the interaction with existing taxes. Because environmental taxes raise revenue which can be used to reduce existing tax distortions, they may present an opportunity to simultaneously improve environmental quality and increase the efficiency of the tax system. This is the "double dividend" hypothesis of environmental policy (see e. g. Pearce 1991 for a general discussion of the double dividend proposition). Second, we did not investigate the problem of carbon leakage. Carbon leakage occurs when emission reduction in regions imposing environmental constraints are (partially) offset by increased emissions in non-acting regions. Leakage might provide an efficiency argument for the concession of tax exemptions or grandfathering permits to energy- and export-intensive industries (see e. g. Böhringer and Rutherford 1996a or Böhringer, Ferris, Rutherford 1996). Third, the model only has a crude representation of the labor market. In the light of high unemployment rates in the EU countries, a more sophisticated treatment of labor market mechanisms (such as union bargaining, labor migration, etc.) would be instructive. Fourth, the model has a static structure. The lack of "true" (intertemporal) dynamics makes it less than perfectly suited for the analysis of adjustment of physical and human capital stocks to carbon emission restrictions.

Fifth, apart from theoretical modeling issues, we see a strong need for the harmonization of physical energy flow data with economic input-output data to strengthen the credibility of carbon policy analysis.

Lines of future research should extend the current framework along these dimensions using additional data and standard economic theory.

Part III: Summary and Conclusions

The greenhouse problem is one of the most challenging issues confronting energy and environmental policy worldwide. Not surprisingly it also ranks high on the political agenda in the European Union.

In response to the greenhouse problem, the EU has adopted the target of stabilizing its CO_2 emissions by 2000 at the level of 1990. Several member states have even further reaching aims. Germany, for example, pursues the objective of a 25-percent reduction in its emissions relative to the level in 1990.

As a policy instrument to reduce CO_2, the European Commission has proposed a combined carbon/energy tax. This proposal reflects the common view that environmental taxes are a cost-effective means of emission abatement.

Although the initiative to introduce a carbon/energy tax throughout the EU failed, the Council of Environment Ministers has left it to the member states to introduce such measures on a national basis. The question then arises as to the likely effects of the introduction of a carbon/energy tax in a single country, in comparison with a European-wide tax. In this regard, it is frequently argued that unilateral abatement measures are ineffective and economically harmful, an assertion which deserves close scrutiny.

A second policy instrument which could achieve CO_2 abatement are tradeable emission permits. Tradeable permits can be used either on a national basis, to implement national emission limits, or on a European basis. In the latter case, this would allow for exploiting interregional differences in the costs of CO_2 abatement, leading to a further reduction of overall costs.

No matter which form European CO_2 abatement policies take, they should be viewed against the background of another major political project of the EU: the completion of the Internal Energy Market. The effect of the Internal Market on CO_2 is particularly important with respect to electricity because of the large share of electricity generation in European CO_2 emissions.

There are several ways in which the Internal Electricity Market may affect CO_2 emissions.

First, increased competitive pressure may lead to increased cost efficiency in the electricity supply industry, implying the selection of technologies with higher thermal efficiency.

Second, increased market risk may increase the preference for less capital intensive power generation technologies. Because these technologies are typically gas-based, the carbon intensity of power generation could be reduced.

Third, the internal market may make national CO_2 abatement programs less effective because national abatement could be avoided by increasing imports.

Given the inter-relatedness of European CO_2 abatement and the completion of the Internal Market, questions arise concerning the following issues:

- the effect on CO_2 of market opening *per se*,
- the effect on CO_2 of market opening in combination with measures of national energy policy,
- the effect on CO_2 of market opening in combination with measures of environmental policy.

The answers to these questions depend on the structural changes induced by these policy measures, i.e., changes in the technology and fuel mix, sectoral changes within countries, and changes in the division of labor between countries.

These changes, in turn, depend on the structural features characterizing the initial situation (CO_2 intensities of the economies, sectoral structure, factor intensities and demand composition of the various sectors, fuel mix). The structural changes an economy will undergo in response to CO_2 abatement will be the more pronounced the more carbon-intensive the fuel mix and the higher the share of energy-producing or energy-intensive sectors in overall output. The structure of the various national electricity supply systems, is particular importance.

Logically, it follows from these considerations that an assessment of market opening and of CO_2 abatement in the European Union requires a sectorally-differentiated multi-country model with a detailed representation of the European electricity supply system.

Before summarizing our conclusions, we would like to point out that the *precise numerical results* depend to a considerable degree on certain basic assumptions. Especially, the base case development is largely shaped by energy price assumptions and the effect of market opening by the assumptions on cost convergence. In spite of these assumptions, we consider our qualitative results and the major orders of magnitude to be rather robust. For instance, it is plausible that electricity imports induced by market opening will be more pronounced in the intermediate term (to the degree to which cost differences persist) than in the long term (when cost differences will have eroded). Also, the general effect of both market opening and CO_2 abatement policies on the fuel mix seems to be undisputable.

With these caveats in mind, the results of the study can be summarized as follows.

The *opening of national electricity markets* leads to a significant relocation of power generation in the various regions and to a change in the fuel mix. Power generation is relocated from Germany, Italy and Spain/Portugal to France, the U.K. and Belgium/Netherlands. In the long term (2020), power generation may decrease by up to 5 percent in Germany and rise by somewhat more than 11 percent in Belgium/Netherlands. The import shares arising from a scenario of unrestricted electricity trade stay well below the limits implied by the European Commission's Internal Market directive. They do not necessarily increase over time because competitive pressure present in the Internal Market leads to a partial convergence of costs, which reduces the incentives for trade.

The effect on electricity-related CO_2 emissions is different in the intermediate term and in the long term. In the intermediate term (2005), there is a relocation of emissions across the various regions in line with the relocation of power generation. Overall, electricity-related emissions increase by almost 3 percent. In the long term (2020), a substantial change in the electricity industry's fuel mix takes place. A strong substitution of gas for coal occurs. As a result of this, CO_2 decreases not only in those countries in which electricity generation decreases, but in the others as well. Overall, electricity-related CO_2 drops by almost 7 percent.

With respect to the general economy, the opening of electricity markets leads to changes in GDP which mirror the changes in power generation: this entails increases in France, the U.K. and Belgium/Netherlands and reductions in Germany, Italy and Spain/Portugal. However, the size of the changes is rather modest. In general, the long-term changes are below 0.2 percent.

With respect to CO_2, intermediate-term increases in all regions are recognizable, totalling 2.5 percent for the EU. In the long term, emissions rise only in France and the U.K., and overall European emissions drop by 0.2 percent.

In the case of a *German nuclear phase-out*, the resulting gap is partly filled by imports and partly by increasing fossil-based domestic power generation. As a (direct or indirect) result of larger German imports, power generation in all European regions increases. The increase in power generation is mainly based on gas. In Germany, gas-based power generation increases by almost 250 percent in the intermediate term. In addition, there is an increase in coal-based electricity by more than 10 percent. Electricity-related CO_2 increases by 11 percent in Germany and by almost 8 percent across the EU. In the long term, the substitution of imports for German power generation is partly reversed. An increase in lignite-based electricity takes place in Germany, and electricity-related CO_2 rises by 19 percent.

With respect to other industries, the nuclear phase-out leads to an increase in the German oil, gas and equipment goods industry. In the long term, there is also an increase in the lignite industry. Most other sectors experience a loss. German GDP decreases by about 0.6 percent in the intermediate term and 0.3 percent in the long term. Total German CO_2 emissions increase by about 2 percent in the intermediate term and 3.5 percent in the long term. Total European CO_2 emissions rise by 0.7 and about 1.5 percent in the intermediate and long term, respectively.

Overall, it can be concluded that both the opening of European electricity markets and a hypothetical German nuclear phase-out will have visible effects on European CO_2 emissions. While the nuclear phase-out will lead to more CO_2, as one would have expected, market opening will trigger a long-term substitution of gas for the more carbon-intensive fuels, such that CO_2 may be expected to be reduced, albeit by a relatively small amount.

The effect of a *carbon/energy tax* designed along the lines proposed by the European Commission in 1992 depends, obviously, on whether the tax is introduced throughout the EU or in a single country. In addition, the effect on the general economy also differs depending on the way in which the tax revenue is recycled into the economy.

The first-order effect of the carbon/energy tax is an increase in the prices of the energy carriers and in the average energy prices charged to the various industries. The energy price changes induced by the EU-wide tax do not only show a substantial dispersion across the industries in each particular country, but also across countries, for each particular industry. This indicates substantial differences in the carbon intensities of energy inputs both across sectors and countries.

In the electricity supply industry, the tax leads to a changing energy mix and a relocation of power generation in the various countries, while the overall level of power generation is kept constant, by assumption. These structural changes lead to a reduction of EU-wide electricity-related CO_2 emissions by almost 5 percent in the intermediate term (2005) and by 13 percent in the long term (2020).

The effects on CO_2 stemming from electricity generation vary across countries and are different in the intermediate and in the long term. In the intermediate term, not all countries reduce their emissions. This result arises because some countries (those with coal-intensive electricity supply) respond to the tax by increasing their imports, given the inflexibility of existing supply structures. This leads to higher emissions in the other countries, which increase their exports. Typically, these switches go along with an increase in gas-based power generation.

In the long term, restructuring of generation capacities becomes more attractive, whereas imports lose part of their cost advantage as a response to the tax. Therefore, electricity-related CO_2 is reduced in *all* countries. This is mainly achieved through a further increase in gas firing.

With respect to the non-energy industries, it is found that, on the average, the basic materials and the construction industries experience the strongest output reduction. In the former case, this is due to the high energy intensity and in the latter case to the dependence on investment demand. The sectoral effects differ from country to country. In most cases, they are negative, but output expansions occur as well. In general, the output effects are below one percent, both in the intermediate and in the long term.

With respect to the general economy, a reduction in EU-wide CO_2 by about 4 percent (2005) and by almost 8 percent (2020) occurs. This is a significantly smaller relative reduction than in the electricity sector, especially in the long term. Comparing the various countries reveals a considerable dispersion of the effect.

The effect on GDP is a reduction by 0.2 percent in the intermediate term, for the EU as a whole. Across the various countries, the effect ranges from a reduction by 0.3 to an increase by 0.1 percent. In the long term, the average is a reduction of about 0.4 percent, with a range from a 0.7 percent reduction to a one percent increase. Investment is affected more by the tax than is consumption, and imports decrease more than exports. The latter is a straightforward consequence of the high share of energy in European imports. As a further consequence, there is an improvement of the EU's aggregate current account vis a vis the rest of the world, and a revaluation of the EU's currency.

If the carbon/energy tax is introduced unilaterally in Germany, the relocation of power generation is, of course, even more pronounced. In this case power generation in Germany is reduced by more than 28 percent in the intermediate term and by almost 58 percent in the long term. This goes along with a reduction of German electricity-related CO_2 emissions by more than 45 percent in the intermediate term and by more than 88 percent in the long term. Power generation and related CO_2 emissions outside Germany rise sharply. In spite of this, EU-wide electricity-related emissions drop. In the intermediate term, they drop more under unilateral taxation than under joint taxation. This can be explained by the high carbon intensity of German electricity in comparison with electricity imported from Belgium/Netherlands and France.

The effects of the German carbon/energy tax on the non-energy sectors of the economy are not very pronounced. Visible sectoral effects do, of course, arise in Germany. Overall, some industries in Germany fare better and others worse under the unilateral than under the joint tax. In contrast to common reasoning, the export-oriented industries tend to be more favorably affected by the unilateral than by the joint tax, which is partly related to differences in the effect on the exchange rate.

With respect to the general economy, it is found that German CO_2 is reduced about 2 ½ times as much as under joint taxation. This increased effectiveness of the tax (in comparison with joint taxation) stems largely from the electricity sector.

As a result of the relocation of CO_2 intensive activities, emissions increase outside of Germany. In spite of these 'leakage effects' the unilateral German carbon/energy tax achieves a reduction of overall European emissions by about 2 percent.

The effect on German GDP is a reduction by about 0.5 percent in the intermediate term and one percent in the long term. The effect on the other countries' GDP is mostly quite small, except for Belgium/Netherlands, which in the intermediate term is a major gainer in terms of GDP.

These results are based on the assumption that the revenue from the carbon/energy tax is redistributed to private households. Different results arise, if the revenue is used to lower the costs of labor to employers. It is found that such a measure simultaneously achieves a reduction of CO_2 emissions and a significant increase in employment and GDP. In general, the effect on employment and GDP increases and that on CO_2 decreases as the wage elasticity becomes smaller. Only in the case of Belgium/Netherlands, a lower wage elasticity leads to a weaker increase in employment. This result is related to a change in comparative advantage. Because the export-oriented industries in Belgium/Netherlands are less labor intensive than their European competitors (especially Germany and the

U.K.), the former benefit less from the reduction of labor costs than the latter. As a consequence of this, exports from Belgium/Netherlands drop, whereas those from Germany and the U.K. rise.

A general insight from these simulations is that the effectiveness of the carbon/energy tax with respect to overall CO_2 emissions in each particular region and in the EU as a whole is largely determined by the structure of the various electricity supply systems. In countries in which rigidities in the structure of the electricity system prevent a short-term change in the fuel mix the carbon/energy tax induces increasing electricity imports. This is possible, because (within each load block) electricity is a homogeneous good. Therefore, trade is restricted only by transmission costs (transmission losses). It is partly for this reason that the European electricity supply system turns out to be more flexible with respect to CO_2 abatement than the rest of the economy. This manifests itself through larger CO_2 reduction rates in the electricity industry than in the general economy. In addition, it should be noted that these reductions are achieved through structural changes alone, not through changes in overall power generation. If these were taken into account, the CO_2 reduction would be even larger.

With respect to unilateral CO_2 reduction measures, it is mainly the electricity industry which contributes to their surprisingly strong effectiveness. Because German electricity generation is much more carbon intensive than electricity generation in countries to which power generation would be shifted in the case of a unilateral German carbon/energy tax, the increase in CO_2 outside Germany is much less than the decrease within Germany. As a consequence, the 'leakage effect' is not too pronounced even in spite of a strong substitution of imports for domestic power generation, and the effect of the unilateral tax on overall European CO_2 emissions is surprisingly strong.

These results indicate that a detailed modeling of the electricity supply system is of considerable importance for a careful analysis of carbon/energy taxes in the context of the Internal Energy Market.

The simulations of permit policies to achieve given CO2 reduction targets suggest that carbon abatement imposes significant economic costs independent of coordinated or uncoordinated action. This result is not surprising given the level of reduction targets (10 %, 20 % and 30 %) and the assumption of a first-best benchmark economy without prior distortions. Abstracting from environmental benefits and initial market distortions such as taxes, rigid wages etc., a restriction of carbon or fossil energy use necessarily makes the economy worse off unless there are zero-cost substitution possibilities. Unilateral EU abatement provides an option to shift part of the abatement costs to the non-abating rest of the world, but gains in the composite ToT (i.e. international substitution effects and income effects) are by far lower than the losses associated with the domestic substitution effect.

In accordance with standard economic theory, the empirical results show that coordination of abatement among EU countries is less costly than uncoordinated action. However, the quantitative results indicate that the benefits from coordination are rather low, i.e., differences in marginal costs of abatement across countries are not large enough to yield significant gains from equalization of these costs through international permit trade.

At the country level, coordinated action does not necessarily imply welfare gains as compared to uncoordinated action. Changes in the domestic substitution effect on the one hand and the composite ToT as well as permit revenue effects on the other hand have opposite welfare implications. Depending on the benchmark economic structure and exogenous elasticities, this may lead to winners and losers from coordination. With winners and losers, coordination is not a no-regret strategy for all countries unless side-payments assure that each country is better off under cooperation than without international cooperation. This calls for negotiation on compensating mechanisms, which might involve significant transaction costs. In view of the small overall benefits from coordination, advocates of cooperation should be aware that these gains might not pay for transaction costs of negotiation or even the implementation costs of a common permit market.

Future research on CO_2 abatement within the EU should address potentially important questions which have not yet been incorporated in our analytical framework such as:

(1) Environmental tax reform: The debate on environmental tax reforms has addressed the question of whether the trade-off between environmental benefits and gross economic costs of environmental taxes prevails in economies where distortionary taxes finance public spending. On the one hand, environmental taxes (permits) raise revenue, which can be used to reduce existing tax distortions. Revenue recycling may present an opportunity to simultaneously improve environmental quality and increase the efficiency of the tax system (the so-called "double dividend" hypothesis of environmental policy). On the other hand, environmental taxes cause distortions on markets for dirty goods and induce spill-over effects to other markets with responses from these markets feeding back. These tax interaction effects are typically negative (see Bovenberg and Goulder 1994). Depending on the nature of the existing tax distortions (level of tax rates or the broadness of the tax base to yield a given tax revenue, demand and supply elasticities of taxed goods etc.) the overall effect of a revenue-neutral tax reform may be negative or positive (see Böhringer, Pahlke and Rutherford 1996 for an empirical analysis). In this context, the incorporation of the major taxes applied in different EU countries would represent a substantial extension to the modeling framework employed for this study.

(2) Carbon leakage: Emission reduction in EU countries may be offset by increased emissions outside the EU due to the relocation of carbon-intensive industries or large cut-backs of EU energy demand which could induce a drop in world energy prices and increase energy demand in other regions. The analysis of carbon leakage calls for a regional extension of the current

modeling framework accounting for other important carbon emitters such as the U.S., Japan, C.I.S. etc. through explicit production and consumption structures as well as bilateral trade links. An extension of the current framework to account for initial non-price competition would be very desirable.

(3) Labor market economics: In the light of the high unemployment rates in EU countries, politicians are concerned about negative implications of carbon abatement on economic performance and the level of employment. The models employed for this study only have a crude representation of the labor market. To address the issue of employment effects in further detail, it would be instructive to add more sophisticated labor market mechanisms such as union bargaining (using game theory), adjustment costs of human capital due to limited mobility, etc.

(4) Reconciliation of physical energy flow data and economic input-output data: The quantitative economic effects of carbon abatement measures depend on the differences in carbon/energy-intensities of sectors within and across regions. In the process of constructing a consistent benchmark data set, it is therefore crucial to employ good estimates on energy and emission flows associated with benchmark monetary flows. The complementary use of physical flow data and economic input-output data reveals severe consistency problems and a thorough reconciliation of both data sources would be an important, yet very tedious task.

References

Armington, P.S. (1969), A Theory of Demand for Products Distinguished by Place of Production, *IMF Staff Papers* 16, 159-178.

Bovenberg, A.L. and L.H. Goulder (1994), Integrating Environmental and Distortionary Taxes: General Equilibrium Analysis, working paper, Stanford University.

Böhringer, C. (1996a), NEWAGE - an integrated modeling system for energy and environmental policy analysis, in: Molt, S. und Fahl, U. (Hrsg.): *Energiemodelle in Deutschland - Stand der Entwicklung*, forthcoming 1997.

Böhringer,C. (1996b), *Allgemeine Gleichgewichtsmodelle als Instrument der energie- und umweltpolitischen Analyse. Theoretische Grundlagen und empirische Anwendung,* Europäischer Verlag der Wissenschaften, Peter Lang.

Böhringer, C. (1996c), The Synthesis of Bottom-Up and Top-Down in Energy Policy Modeling, Institut für Energiewirtschaft und Rationelle Energieanwendung, Universität Stuttgart.

Böhringer, C. (1996d), Fossil Fuels Subsidies and Environmental Constraints, *Environmental and Resource Economics* 8.

Böhringer, C. and T.F. Rutherford (1996a), Carbon Taxes with Exemptions in an Open Economy - A General Equilibrium Analysis of the German Tax Initiative, forthcoming: *Journal of Environmental Economics and Management 32-2,* 1997.

Böhringer, C. and T.F. Rutherford (1996b), The Efficiency Costs of Voluntary Agreements in Environmental Policy: An Analytical Framework, Institut für Energiewirtschaft und Rationelle Energieanwendung, Universität Stuttgart, Deparment of Economics, University of Colorado.

Böhringer, C, Ferris, M. and T. Rutherford (1996), Alternative CO_2 abatement strategies for the European Union, in: *Economic aspects of Environmental policy making in a federal state*, Edward Elgar Publishing, forthcoming 1997.

Böhringer, C., Pahlke, A. and T.F.Rutherford (1996), Environmental Tax Reforms and the Prospects for a Double-Dividend - an Intertemporal CGE Analysis for Germany, Institut für Energiewirtschaft und Rationelle Energieanwendung, Universität Stuttgart.

Burniaux, J.-M., Martin, J.P., Nicoletti, G., Oliveira Martins, J. (1992), GREEN - A Multi-Sector, Multi-Region General Equilibrium Model for Quantifying the Costs of Curbing CO_2 Emissions: A Technical Manual, *OECD Working Paper* No. 116, Paris.

Conrad, K. (1993), Problems of Modeling CO_2 Reduction with CGE Models, in: W. Kuckshinrichs, W. Pfaffenberger, W. Ströbele (eds.), *Economics of the Greenhouse Effect*, Jülich.

Dirkse, S. and M. Ferris (1995), The PATH Solver: A Non-Monotone Stabilization Scheme for Mixed Complementarity Problems, *Optimization Methods & Software*.

EUROSTAT (1991), *National Accounts ESVG: Detailed Tables by Branches 1980-1988*, Luxembourg.

EUROSTAT (1992), *National Accounts ESVG: Input -Output Tables 1985*, Luxembourg.

Fehr, H., Rosenberg, C. and W. Wiegard (1994), *VAT Harmonization in the EU*, Berlin, Heidelberg, New York: Springer.

Gerstenberger, W., Heinze, J., Hummel, M., Vogler-Ludwig, K. (1989), *Sektorale Kapitalbildung in der deutschen Wirtschaft nach dem Eigentümer- und Benutzerkonzept*, ifo-Studien zur Strukturforschung, München.

Goulder, L.H. (1994), Environmental Taxation and the Double Dividend: A Reader's Guide, *mimeo*, Stanford: Stanford University.

Hoster, F. (1996a), *Auswirkungen des europäischen Binnenmarktes für Energie auf die deutsche Elektrizitätswirtschaft - Ein Ansatz zur Analyse ordnungs- und umweltpolitischer Instrumente in der Elektrizitätswirtschaft*, München: Oldenbourg, (Schriften des Energiewirtschaftlichen Instituts, Bd. 49).

Hoster, F. (1996b), Auswirkungen des Europäischen Binnenmarktes für Strom auf Stromhandel und Erzeugungsstruktur, *Zeitschrift für Energiewirtschaft (ZfE), 20.Jg. Heft 4.*

Hoster, F. (1997), The Modalities of Grid Access - Toward an Integrated European Grid ?, with C. Riechmann, W. Schulz, *Economie et Société*, forthcoming 1997.

IEA (1994), *Climate Change Policy Initiatives, 1994 Update*, Paris.

IEA (1996), *Energy Prices and Taxes*, Paris.

Koopmans, T.C. (1951), *Activity Analysis of Production and Allocation*, New York.

Manne, A.S. (1981), *ETA-MACRO: A User's Guide*, EA-1724, Electric Power Research Institute, Palo Alto, CA.

Manne, A.S., Richels, R.G. (1992), *Buying Greenhouse Insurance: The Economic Costs of CO_2 Emission Limits*, Cambridge, MIT Press.

Pearce, D.W. (1991), The Role of Carbon Taxes in Adjusting to Global Warming, *Economic Journal* 101, 938-948.

PROGNOS (1995), *Energiereport 2020*, Stuttgart.

Rutherford, T. (1992), Sequential Joint Maximization, working paper 92-8, Department of Economics, University of Colorado.

Rutherford, T. (1993), MILES: A Mixed Inequality and nonLinear Equation Solver, *mimeo*, Department of Economics, University of Colorado.

Rutherford, T. (1994), Extensions of GAMS for Complementarity Problems arising in applied Economic Analysis, *mimeo*, Department of Economics, University of Colorado.

Shoven, J.B. and J. Whalley (1992), *Applying General Equilibrium*, Amsterdam

Takayama, T. and G.G. Judge (1971), *Spatial and Temporal Price and Allocation Models*, Amsterdam

Turvey, R., Anderson, D., (1977), *Electricity Economics: Essays and Case Studies*, Washington, DC.

UNIPEDE (1995), *EURPROG 1995: Programmes and Prospects for the Electricity Sector*, 23rd Edition, Paris.

WEFA (1995), *Comptes Harmonisés sur les Echanges et L'Economie Mondiale*, Paris.

Welsch, H. (1996a), Recycling of Carbon/Energy Taxes and the Labor Market: A General Equilibrium Analysis for the European Community, *Environmental and Resource Economics* 8, 141-155.

Welsch, H. (1996b), Joint vs. Unilateral Carbon/Energy Taxation in the European Union: Results from a General Equilibrium Model, *Konjunkturpolitik*, forthcoming.

Whalley, J. (1985), *Trade Liberalization among Major World Trading Areas*, Cambridge.

Appendix A: Internal Market for Electricity - Detailed Results

Table A1. Effect of Market Opening in Germany

	2000	2005	2010	2015	2020
Electricity Industry					
Total Generation	0.65	0.88	-0.55	-4.40	-4.74
HC	-0.95	-4.35	-3.11	-15.08	-19.81
BC	3.21	0.00	-0.03	0.00	-18.83
Oil	0.00	179.44	0.00	0.00	3411.11
GAS	0.00	10.15	6.51	1.60	55.81
NUC	0.00	0.00	0.00	0.00	0.00
Others	0.00	0.00	0.00	0.00	0.00
Imports	-3.71	-2.21	6.41	22.91	49.22
Exports	4.57	-13.73	-0.60	-29.20	-6.61
CO_2	1.73	-1.68	-1.15	-6.17	-11.83
Sectoral Output					
Agriculture	0.03	0.03	0.04	0.01	0.15
Hard Coal	0.00	0.00	0.00	0.00	0.00
Lignite	2.17	0.06	0.00	0.00	-8.33
Oil	0.06	1.83	2.88	3.28	5.32
Gas	0.48	0.39	0.42	0.45	3.47
Electricity	0.65	-0.88	-0.55	-4.40	-4.74
Basic Materials	-0.07	-0.15	-0.06	-0.08	0.04
Equipment	0.04	-0.24	-0.18	-0.16	-0.14
Consumption Goods	0.09	-0.04	-0.01	0.03	0.15
Construction	0.23	0.02	-0.01	0.09	-0.04
Transport	0.04	-0.02	0.06	0.05	0.15
Private Services	0.13	-0.01	0.00	0.09	0.21
Public Services	-0.07	-0.03	0.03	0.08	0.12
Macroeconomy					
CO_2	0.28	0.47	1.27	-0.08	-0.99
Hard Coal	-0.45	-2.28	-1.53	-7.66	-7.85
Lignite	2.56	0.07	0.06	0.07	-10.03
Oil	0.15	2.44	3.87	4.21	5.17
Gas	-0.11	0.35	0.59	0.90	0.88
GDP	0.06	-0.12	-0.10	-0.08	-0.12
Consumption	0.46	0.16	-0.06	0.14	0.38
Investment	0.52	0.16	0.02	0.18	-0.21
Export	-0.13	-0.12	0.05	-0.20	-0.13
Import	0.21	-0.22	-0.41	-0.29	0.14
Employment	0.14	-0.10	-0.07	-0.01	0.02
Capital Stock	0.00	-0.02	-0.03	0.00	0.01
Exchange Rate	0.03	0.03	-0.20	-0.33	-0.24

Table A2. Effect of Market Opening in France

	2000	2005	2010	2015	2020
Electricity Industry					
Total Generation	8.12	7.23	5.76	5.72	2.71
HC	203.80	114.68	75.90	43.82	17.80
BC	0.00	0.00	0.00	0.00	0.00
Oil	0.00	0.00	0.00	0.00	873.08
GAS	0.00	133.75	348.61	1582.36	2.23
NUC	0.00	0.00	0.00	0.00	0.00
Others	6.20	2.60	0.52	4.83	5.29
Imports	1235.94	896.39	308.42	399.01	220.95
Exports	76.70	75.74	54.82	49.14	27.33
CO_2	186.14	133.67	86.26	57.86	26.98
Sectoral Output					
Agriculture	0.05	0.15	0.08	0.03	0.09
Hard Coal	22.89	17.59	7.80	2.59	-1.52
Lignite	0.00	0.00	0.00	0.00	0.00
Oil	-0.41	0.83	1.46	1.65	2.46
Gas	-0.16	0.42	0.51	1.12	0.86
Electricity	8.12	7.23	5.76	5.72	2.71
Basic Materials	0.39	0.33	-0.05	-0.32	-0.27
Equipment	-0.36	0.01	-0.42	-0.69	-0.61
Consumption Goods	0.06	0.19	-0.04	-0.18	-0.05
Construction	-1.18	0.36	0.19	0.22	0.28
Transport	-0.03	0.33	0.15	0.05	0.16
Private Services	-0.17	0.29	0.15	0.14	0.31
Public Services	0.11	0.17	0.09	0.03	0.12
Macroeconomy					
CO_2	7.96	8.26	7.37	6.39	4.58
Hard Coal	60.05	46.00	32.83	21.55	10.65
Lignite	0.00	0.00	0.00	0.00	0.00
Oil	-0.47	1.83	3.04	3.18	4.22
Gas	0.65	1.62	1.62	3.65	1.21
GDP	-0.08	0.24	0.03	-0.03	0.05
Consumption	-0.57	0.51	0.47	0.66	1.02
Investment	-2.40	0.42	0.50	0.74	0.78
Export	1.02	0.51	0.21	-0.20	-0.44
Import	-0.13	0.32	-0.04	-0.05	0.08
Employment	-0.33	0.28	-0.02	-0.12	0.03
Capital Stock	0.00	-0.03	0.00	0.04	0.09
Exchange Rate	0.03	0.03	-0.20	-0.33	-0.24

Table A3. Effect of Market Opening in U.K.

	2000	2005	2010	2015	2020
Electricity Industry					
Total Generation	6.69	7.28	5.70	3.25	2.50
HC	32.50	10.80	-2.46	-5.09	-8.58
BC	0.00	0.00	0.00	0.00	0.00
Oil	4.91	33.16	7.14	0.00	0.00
GAS	2.91	4.69	22.82	22.03	26.67
NUC	0.00	0.00	0.00	0.00	0.00
Others	0.00	0.00	0.00	0.00	0.00
Imports	-90.87	-100.00	-87.66	-61.47	-57.63
Exports	0.00	0.00	0.00	0.00	0.00
CO_2	15.89	13.51	4.10	0.83	-1.30
Sectoral Output					
Agriculture	-0.04	0.31	0.34	0.31	0.31
Hard Coal	13.69	5.83	-2.51	-4.51	-6.70
Lignite	0.00	0.00	0.00	0.00	0.00
Oil	-0.01	-0.62	-1.38	-1.60	-1.68
Gas	-0.89	0.72	1.46	1.40	1.79
Electricity	6.69	7.28	5.70	3.25	2.50
Basic Materials	0.18	0.31	0.34	0.27	0.11
Equipment	-0.64	0.34	0.44	0.18	0.05
Consumption Goods	-0.30	0.33	0.31	0.22	0.21
Construction	-2.48	0.71	0.65	0.29	0.66
Transport	-0.35	0.30	0.27	0.16	0.15
Private Services	-0.55	0.30	0.22	0.14	0.29
Public Services	-0.19	0.29	0.11	-0.01	0.05
Macroeconomy					
CO_2	2.13	3.39	1.93	1.25	0.95
Hard Coal	17.23	8.42	-1.45	-3.64	-6.04
Lignite	0.00	0.00	0.00	0.00	0.00
Oil	-0.46	3.95	2.91	2.56	2.49
Gas	-0.38	1.07	2.82	2.46	2.89
GDP	-0.41	0.27	0.18	0.04	0.10
Consumption	-1.53	0.16	-0.19	-0.27	0.42
Investment	-4.84	0.96	0.83	0.11	0.96
Export	0.55	-0.71	-1.07	-1.16	-1.41
Import	-1.05	0.00	-0.47	-0.73	-0.44
Employment	-0.84	0.35	0.19	-0.03	0.03
Capital Stock	-0.01	0.06	0.19	0.29	0.35
Exchange Rate	0.03	0.03	-0.20	-0.33	-0.24

Table A4. Effect of Market Opening in Italy

	2000	2005	2010	2015	2020
Electricity Industry					
Total Generation	-17.01	-19.13	-15.45	-9.79	-1.23
HC	-5.76	-0.62	-3.52	-34.08	-38.91
BC	0.00	29.13	29.13	29.13	0.00
Oil	-38.39	-13.61	-8.48	16.54	103.96
GAS	-4.98	-72.86	-33.83	1.94	22.65
NUC	0.00	0.00	0.00	0.00	0.00
Others	3.01	1.81	-0.37	-1.20	1.78
Imports	100.40	148.32	137.94	95.16	17.98
Exports	0.00	0.00	-84.25	630.23	2847.78
CO_2	-21.65	-14.72	-13.15	-16.07	-9.51
Sectoral Output					
Agriculture	-0.08	0.07	0.11	0.13	0.21
Hard Coal	-3.99	10.12	9.61	-19.04	-26.84
Lignite	0.00	0.00	0.00	0.00	0.00
Oil	-1.04	2.12	4.37	5.22	6.51
Gas	0.28	-11.85	-9.62	-0.53	4.58
Electricity	-17.01	-19.13	-15.45	-9.79	-1.23
Basic Materials	-0.47	0.38	0.27	-0.28	0.06
Equipment	0.37	-0.28	-0.33	-0.07	-0.04
Consumption Goods	-0.06	0.18	0.16	0.16	0.32
Construction	1.71	-0.25	0.19	-0.05	-0.22
Transport	0.13	0.09	0.18	0.17	0.20
Private Services	0.27	0.10	0.33	0.25	0.27
Public Services	-0.23	-0.02	0.37	0.59	0.32
Macroeconomy					
CO_2	-5.13	1.89	2.24	-2.51	-2.67
Hard Coal	-3.73	-2.72	-6.08	-20.18	-25.39
Lignite	0.00	32.70	32.70	32.48	0.00
Oil	-7.61	0.62	4.22	6.44	11.17
Gas	0.84	10.97	6.19	-2.44	-4.90
GDP	0.04	0.15	0.38	-0.03	-0.04
Consumption	1.37	0.28	1.00	0.53	0.49
Investment	4.67	-1.04	0.18	-0.12	-0.97
Export	-1.13	0.39	-0.19	-0.10	0.36
Import	1.45	-0.42	-0.67	-0.13	-0.04
Employment	0.59	-0.35	-0.23	0.02	0.08
Capital Stock	0.00	-0.02	-0.02	0.08	0.10
Exchange Rate	0.03	0.03	-0.20	-0.33	-0.24

Table A5. Effect of Market Opening in Belgium/Netherlands

	2000	2005	2010	2015	2020
Electricity Industry					
Total Generation	-9.16	1.62	3.76	11.12	11.40
HC	1.54	-4.91	-2.22	0.00	-50.49
BC	0.00	0.00	0.00	0.00	0.00
Oil	0.00	0.00	7.72	0.00	0.00
GAS	-43.63	23.85	13.16	30.97	129.50
NUC	0.00	-7.17	0.00	0.00	0.00
Others	0.00	0.00	0.00	0.00	0.00
Imports	55.07	-1.24	8.42	-29.41	11.74
Exports	-15.19	45.99	118.71	206.46	342.47
CO_2	-6.71	0.31	1.59	9.13	-10.75
Sectoral Output					
Agriculture	-0.03	0.08	-0.02	-0.12	-0.20
Hard Coal	0.25	-5.12	-7.58	-8.03	-20.84
Lignite	0.00	0.00	0.00	0.00	0.00
Oil	-0.20	3.05	5.15	5.32	6.17
Gas	-0.15	-0.04	-0.13	0.31	1.45
Electricity	-9.16	1.62	3.76	11.12	11.40
Basic Materials	-0.12	0.02	-0.23	-0.46	-0.72
Equipment	0.25	-0.18	-0.35	-0.83	-0.96
Consumption Goods	0.08	0.09	-0.04	-0.28	-0.30
Construction	0.51	-0.21	0.10	-0.16	-0.13
Transport	0.04	0.05	-0.09	-0.32	-0.31
Private Services	0.15	0.07	0.07	0.06	0.40
Public Services	-0.09	0.05	0.10	0.13	0.48
Macroeconomy					
CO_2	-0.57	0.98	1.84	2.25	-2.87
Hard Coal	1.20	-3.13	-1.33	-0.03	-29.33
Lignite	0.00	0.00	0.00	0.00	0.00
Oil	0.49	2.68	4.42	4.27	4.88
Gas	-2.35	1.89	1.63	1.96	5.00
GDP	0.00	-0.01	-0.04	-0.05	0.21
Consumption	0.41	0.26	0.39	0.67	1.55
Investment	1.20	-0.54	0.35	-0.05	-0.26
Export	-0.08	0.17	0.20	0.10	0.21
Import	0.26	-0.10	-0.05	-0.16	-0.03
Employment	0.25	-0.03	-0.02	-0.11	0.04
Capital Stock	0.03	-0.03	-0.07	-0.09	-0.07
Exchange Rate	0.03	0.03	-0.20	-0.33	-0.24

Table A6. Effect of Market Opening in Spain/Portugal

	2000	2005	2010	2015	2020
Electricity Industry					
Total Generation	-4.33	-2.84	-1.64	-1.06	-1.30
HC	-4.94	-3.13	-5.36	-2.70	-24.30
BC	0.00	0.00	0.00	0.00	0.00
Oil	0.00	0.00	0.00	0.00	0.00
GAS	0.00	4.15	-1.14	-1.42	16.39
NUC	-8.96	-8.99	0.00	0.00	0.00
Others	-1.47	-4.80	-0.50	0.32	2.71
Imports	214.45	151.91	91.53	75.85	84.51
Exports	0.00	9.04	0.00	174.02	79.41
CO_2	-4.08	-1.98	-3.67	-1.62	-7.54
Sectoral Output					
Agriculture	-0.03	0.01	-0.08	0.04	-0.28
Hard Coal	-1.30	-3.32	-5.20	-2.68	-8.33
Lignite	0.00	0.00	0.00	-0.25	-0.34
Oil	-0.12	2.03	3.38	3.70	6.87
Gas	0.31	1.08	-1.44	-1.64	9.26
Electricity	-4.33	-2.84	-1.64	-1.06	-1.30
Basic Materials	-0.24	-0.27	-0.33	0.23	0.34
Equipment	0.14	-0.21	-0.46	-0.28	0.71
Consumption Goods	0.00	-0.03	-0.09	0.13	-0.45
Construction	0.79	0.10	-0.19	-1.26	-1.01
Transport	-0.01	0.11	0.14	0.37	0.07
Private Services	0.11	0.04	0.07	0.04	-1.57
Public Services	-0.11	-0.12	-0.01	-0.22	-1.59
Macroeconomy					
CO_2	-1.02	0.57	0.98	1.59	-1.23
Hard Coal	-3.32	-2.11	-3.68	-1.37	-13.18
Lignite	0.03	0.03	0.08	-0.19	-0.98
Oil	0.07	2.27	3.78	3.78	3.36
Gas	-0.19	-1.15	-1.61	-1.65	6.38
GDP	-0.01	-0.20	-0.25	-0.21	-1.80
Consumption	0.53	0.36	0.49	-0.28	-5.14
Investment	2.01	0.46	-0.09	-3.07	-1.47
Export	-0.41	-0.12	-0.20	0.66	2.34
Import	0.70	-0.11	-0.78	-1.32	0.44
Employment	0.21	-0.10	-0.19	-0.15	0.07
Capital Stock	-0.03	-0.06	-0.14	-0.16	-0.12
Exchange Rate	0.03	0.03	-0.20	-0.33	-0.24

Table A7. Effect of Market Opening in the EU

	2000	2005	2010	2015	2020
Electricity Industry					
Total Generation	0.12	0.12	0.09	0.06	0.05
HC	10.65	6.48	2.20	-8.62	-21.93
BC	2.89	0.02	0.35	0.11	-16.81
Oil	-22.33	-1.45	-3.55	4.61	42.21
GAS	-4.63	-7.98	-1.65	11.28	33.45
NUC	-0.68	-1.09	-0.30	0.00	0.00
Others	-0.35	-0.07	-0.24	1.43	2.30
Imports	57.19	62.12	53.25	33.97	34.51
Exports	58.97	64.74	55.15	34.80	35.38
CO_2	2.81	2.76	0.88	-2.11	-6.89
Sectoral Output					
Agriculture	-0.01	0.11	0.09	0.07	0.08
Hard Coal	2.89	2.78	-0.33	-4.85	-9.91
Lignite	1.61	0.04	0.00	-0.04	-6.57
Oil	-0.24	0.82	1.21	1.20	1.85
Gas	-0.19	-0.68	-0.43	0.61	2.57
Electricity	0.12	0.12	0.09	0.06	0.05
Basic Materials	-0.06	0.09	0.02	-0.11	-0.03
Equipment	-0.07	-0.10	-0.17	-0.22	-0.16
Consumption Goods	-0.02	0.12	0.07	0.04	0.07
Construction	-0.25	0.15	0.20	0.00	0.04
Transport	-0.04	0.15	0.14	0.09	0.11
Private Services	-0.02	0.14	0.14	0.13	0.14
Public Services	-0.09	0.10	0.11	0.09	0.07
Macroeconomy					
CO_2	0.68	2.49	2.46	1.21	-0.21
Hard Coal	6.70	3.95	0.96	-4.33	-11.73
Lignite	2.22	0.10	0.09	0.07	-8.95
Oil	-1.60	2.23	3.63	4.03	5.25
Gas	-0.40	2.16	2.23	1.54	1.92
GDP	-0.07	0.08	0.06	-0.05	-0.13
Consumption	0.00	0.28	0.30	0.26	0.24
Investment	-0.50	0.15	0.34	0.02	0.04
Export	0.03	-0.01	-0.15	-0.25	-0.13
Import	0.10	-0.08	-0.35	-0.35	-0.03
Employment	-0.08	0.04	-0.03	-0.05	0.04
Capital Stock	0.00	-0.01	0.01	0.06	0.10
Exchange Rate	0.03	0.03	-0.20	-0.33	-0.24

Appendix B: Nuclear Phase Out in Germany - Detailed Results

Table B1. Effects of Nuclear Phase Out in Germany

	2000	2005	2010	2015	2020
Electricity Industry					
Total Generation	-10.00	-17.22	-15.49	-14.16	-13.45
HC	8.82	11.32	13.15	12.36	14.36
BC	0.00	-0.05	0.00	0.00	23.20
Oil	0.00	644.93	1937.78	0.00	0.00
GAS	0.00	241.10	162.29	145.71	39.27
NUC	-40.00	-100.00	-100.00	-100.00	-100.00
Others	0.00	0.00	0.00	0.00	0.00
Imports	77.89	187.63	156.80	114.06	83.46
Exports	-52.02	-49.65	-28.81	-31.45	-31.94
CO_2	4.01	11.01	11.76	11.81	19.59
Sectoral Output					
Agriculture	-0.07	-0.11	-0.06	-0.01	-0.05
Hard Coal	0.00	0.00	0.00	0.00	0.00
Lignite	-0.19	-0.41	-0.39	-0.36	8.83
Oil	0.59	2.16	1.84	0.73	0.73
Gas	0.19	6.60	6.99	7.84	4.26
Electricity	-10.00	-17.22	-15.49	-14.16	-13.45
Basic Materials	-0.26	-0.46	-0.30	-0.15	-0.21
Equipment	0.36	0.66	0.46	0.50	0.36
Consumption Goods	0.09	0.15	0.20	0.26	0.24
Construction	0.30	-0.15	-0.40	-0.27	-0.29
Transport	-0.04	-0.09	-0.08	-0.01	-0.03
Private Services	0.02	-0.17	-0.21	-0.15	-0.14
Public Services	-0.03	-0.06	-0.05	0.03	0.03
Macroeconomy					
CO_2	1.76	2.31	2.51	2.17	3.46
Hard Coal	5.43	7.74	8.49	8.10	6.60
Lignite	-0.19	-0.49	-0.48	-0.45	10.87
Oil	0.70	1.74	1.41	0.81	0.85
Gas	0.40	-1.38	-0.27	0.24	1.09
GDP	-0.10	-0.55	-0.53	-0.42	-0.32
Consumption	0.10	-0.48	-0.54	-0.50	-0.35
Investment	0.69	-0.14	-0.90	-0.89	-0.88
Export	-0.15	0.21	0.36	0.35	0.29
Import	0.30	0.65	0.44	0.33	0.20
Employment	0.12	-0.01	-0.09	0.02	0.04
Capital Stock	0.06	0.07	-0.04	-0.09	-0.10
Exchange Rate	0.02	0.03	0.03	0.03	-0.02

Table B2. Effects of Nuclear Phase Out in France

	2000	2005	2010	2015	2020
Electricity Industry					
Total Generation	0.64	3.28	1.64	0.96	0.69
HC	6.32	3.62	2.22	4.74	0.82
BC	0.00	0.00	0.00	0.00	0.00
Oil	0.00	1181.79	0.00	0.00	13.23
GAS	0.00	718.42	177.18	19.70	31.80
NUC	0.00	0.00	0.00	0.00	0.00
Others	-0.14	0.46	0.67	0.42	-0.16
Imports	18.52	11.53	20.47	36.53	86.04
Exports	4.87	16.52	10.54	7.35	10.97
CO_2	6.43	16.36	8.92	6.43	3.56
Sectoral Output					
Agriculture	0.01	0.03	0.05	0.01	0.03
Hard Coal	1.86	2.53	3.23	2.07	0.99
Lignite	0.00	0.00	0.00	0.00	0.00
Oil	-0.03	0.40	0.02	-0.04	0.02
Gas	-0.03	0.64	0.44	0.14	0.16
Electricity	0.64	3.28	1.64	0.96	0.69
Basic Materials	0.03	0.33	0.22	0.02	0.00
Equipment	-0.03	-0.02	0.06	-0.15	-0.14
Consumption Goods	0.00	0.01	0.05	-0.05	-0.02
Construction	-0.04	-0.33	0.20	0.01	0.00
Transport	-0.01	0.00	0.09	-0.03	0.00
Private Services	0.00	-0.08	0.09	0.00	0.01
Public Services	-0.01	0.01	0.01	-0.02	0.00
Macroeconomy					
CO_2	0.73	1.54	0.91	0.67	0.31
Hard Coal	3.74	1.98	2.73	2.96	0.70
Lignite	0.00	0.00	0.00	0.00	0.00
Oil	-0.04	0.79	-0.02	-0.05	0.05
Gas	-0.01	3.18	1.77	0.57	0.67
GDP	0.00	0.01	0.11	0.00	0.01
Consumption	-0.05	-0.50	-0.02	-0.05	-0.06
Investment	-0.10	-0.83	0.16	-0.03	-0.09
Export	-0.04	0.19	-0.16	-0.26	-0.18
Import	-0.02	-0.25	-0.01	-0.10	-0.08
Employment	-0.02	-0.08	0.13	-0.04	-0.03
Capital Stock	0.00	0.00	0.00	0.02	0.03
Exchange Rate	0.02	0.03	0.03	0.03	-0.02

Table B3. Effects of Nuclear Phase Out in U.K.

	2000	2005	2010	2015	2020
Electricity Industry					
Total Generation	0.96	0.54	0.84	0.95	1.75
HC	3.63	0.28	-0.57	0.13	0.00
BC	0.00	0.00	0.00	0.00	0.00
Oil	0.00	0.00	-8.53	0.00	0.00
GAS	0.68	1.22	4.31	3.18	6.11
NUC	0.00	0.00	0.00	0.00	0.00
Others	0.00	0.00	0.00	0.00	0.00
Imports	-100.00	0.00	-78.76	-63.84	-55.74
Exports	20.01	14.12	20.78	13.38	116.63
CO_2	1.95	0.56	0.09	0.80	1.35
Sectoral Output					
Agriculture	-0.01	0.08	0.05	0.01	0.03
Hard Coal	1.71	0.76	-0.04	0.18	0.09
Lignite	0.00	0.00	0.00	0.00	0.00
Oil	0.03	0.10	0.07	0.05	0.06
Gas	-0.13	-0.01	0.20	0.14	0.15
Electricity	0.96	0.54	0.84	0.95	1.75
Basic Materials	0.02	0.23	0.11	0.01	-0.01
Equipment	-0.13	0.13	-0.01	-0.14	-0.15
Consumption Goods	-0.06	0.09	0.03	-0.03	-0.03
Construction	-0.38	-0.08	0.12	0.11	0.09
Transport	-0.07	0.06	0.04	-0.01	0.00
Private Services	-0.09	0.02	0.06	0.03	0.02
Public Services	-0.04	0.04	0.04	0.02	-0.01
Macroeconomy					
CO_2	0.27	0.06	0.00	0.14	0.19
Hard Coal	2.05	0.22	-0.44	0.09	-0.04
Lignite	0.00	0.00	0.00	0.00	0.00
Oil	-0.15	-0.08	-0.39	-0.05	-0.11
Gas	-0.02	0.11	0.49	0.30	0.49
GDP	-0.07	0.08	0.08	0.03	0.03
Consumption	-0.26	-0.28	-0.01	0.03	-0.02
Investment	-0.76	-0.35	0.11	0.12	0.09
Export	0.01	-0.03	-0.21	-0.28	-0.24
Import	-0.21	-0.20	-0.08	-0.08	-0.09
Employment	-0.14	0.07	0.06	0.00	-0.03
Capital Stock	0.00	-0.01	0.00	0.03	0.05
Exchange Rate	0.02	0.03	0.03	0.03	-0.02

Table B4. Effects of Nuclear Phase Out in Italy

	2000	2005	2010	2015	2020
Electricity Industry					
Total Generation	9.36	12.54	11.45	8.41	4.43
HC	5.92	-2.15	-0.48	0.00	11.45
BC	10.24	0.00	0.00	0.00	0.00
Oil	25.23	-3.05	-2.98	3.55	2.65
GAS	0.00	206.46	39.78	18.48	2.66
NUC	0.00	0.00	0.00	0.00	0.00
Others	-4.00	-2.49	-0.83	1.50	-0.31
Imports	-22.87	-30.69	-36.07	-31.60	-26.94
Exports	0.00	1494.74	663.33	569.43	104.60
CO_2	13.26	5.99	7.49	6.72	6.19
Sectoral Output					
Agriculture	0.04	-0.10	-0.10	-0.12	-0.08
Hard Coal	3.42	-9.39	-10.09	-6.92	3.86
Lignite	0.00	0.00	0.00	0.00	0.00
Oil	0.47	-0.56	-0.70	-0.36	-0.25
Gas	-0.19	10.51	8.24	5.30	1.16
Electricity	9.36	12.54	11.45	8.41	4.43
Basic Materials	0.13	-0.27	-0.27	-0.21	0.04
Equipment	-0.12	-0.06	-0.06	-0.04	-0.20
Consumption Goods	0.02	-0.20	-0.16	-0.14	-0.08
Construction	-0.28	0.08	-0.31	-0.15	-0.16
Transport	-0.01	-0.14	-0.15	-0.13	-0.09
Private Services	-0.01	-0.14	-0.25	-0.22	-0.11
Public Services	0.03	-0.17	-0.29	-0.37	-0.20
Macroeconomy					
CO_2	2.84	-2.18	-1.05	-0.04	1.38
Hard Coal	4.10	0.20	2.74	2.65	6.32
Lignite	11.39	0.00	0.00	0.00	0.00
Oil	3.42	-1.09	-1.11	-0.26	-0.08
Gas	-0.08	-7.53	-4.69	-2.04	0.47
GDP	0.06	-0.28	-0.42	-0.32	-0.08
Consumption	-0.14	-0.30	-0.71	-0.55	-0.26
Investment	-0.84	0.42	-0.58	-0.21	-0.55
Export	0.07	-0.35	-0.03	0.04	-0.07
Import	-0.41	-0.08	-0.10	0.06	-0.30
Employment	-0.10	0.04	0.01	0.03	-0.05
Capital Stock	-0.02	-0.01	-0.04	-0.07	-0.09
Exchange Rate	0.02	0.03	0.03	0.03	-0.02

Table B5. Effects of Nuclear Phase Out in Belgium/Netherlands

	2000	2005	2010	2015	2020
Electricity Industry					
Total Generation	12.17	18.18	18.12	16.89	14.30
HC	0.00	1.75	1.82	0.00	0.00
BC	0.00	0.00	0.00	0.00	0.00
Oil	0.00	0.00	0.00	0.00	0.00
GAS	87.46	58.94	47.52	39.91	26.60
NUC	0.00	0.00	0.00	0.00	0.00
Others	0.00	0.00	0.00	0.00	0.00
Imports	-35.72	11.79	-5.80	-22.20	-7.84
Exports	77.20	437.82	224.34	160.08	98.00
CO_2	10.63	15.88	16.12	14.69	13.67
Sectoral Output					
Agriculture	-0.01	-0.25	-0.29	-0.29	-0.26
Hard Coal	0.25	-3.53	-3.66	-2.91	-2.03
Lignite	0.00	0.00	0.00	0.00	0.00
Oil	-0.20	-0.14	-0.15	-0.16	-0.05
Gas	0.28	0.86	0.85	0.76	0.47
Electricity	12.17	18.18	18.12	16.89	14.30
Basic Materials	-0.06	-0.81	-0.91	-0.78	-0.68
Equipment	-0.33	-0.86	-1.18	-1.04	-0.80
Consumption Goods	-0.12	-0.57	-0.62	-0.53	-0.39
Construction	-0.13	0.34	-0.21	-0.20	-0.06
Transport	-0.09	-0.40	-0.47	-0.42	-0.30
Private Services	0.05	0.14	0.05	0.05	0.12
Public Services	0.07	0.20	0.28	0.25	0.26
Macroeconomy					
CO_2	1.16	1.67	1.66	1.32	1.59
Hard Coal	-0.11	1.18	1.16	0.01	0.02
Lignite	0.00	0.00	0.00	0.00	0.00
Oil	-0.33	0.20	0.15	0.06	0.13
Gas	3.05	2.96	2.87	2.60	2.86
GDP	0.11	0.24	0.15	0.13	0.16
Consumption	0.23	1.07	1.01	0.85	0.82
Investment	-0.26	1.34	0.08	-0.04	0.14
Export	-0.07	0.04	0.01	-0.07	-0.01
Import	-0.19	-0.01	-0.20	-0.24	-0.14
Employment	-0.07	0.06	-0.09	-0.05	0.08
Capital Stock	-0.04	-0.01	-0.03	-0.08	-0.10
Exchange Rate	0.02	0.03	0.03	0.03	-0.02

Table B6. Effects of Nuclear Phase Out in Spain/Portugal

	2000	2005	2010	2015	2020
Electricity Industry					
Total Generation	1.17	1.45	0.97	0.91	1.14
HC	3.17	-1.76	-0.52	0.00	3.05
BC	0.00	0.00	0.00	0.00	0.00
Oil	0.00	0.00	0.00	0.00	0.00
GAS	0.00	16.53	4.51	3.00	1.55
NUC	0.00	0.00	0.00	0.00	0.00
Others	-0.97	0.53	1.39	-0.37	-0.77
Imports	-17.59	-18.52	-22.29	-27.66	-32.93
Exports	0.00	184.82	132.77	16.28	48.22
CO_2	2.56	0.36	0.62	0.83	1.76
Sectoral Output					
Agriculture	0.00	-0.01	-0.01	0.00	0.04
Hard Coal	0.72	-0.42	-0.20	0.29	1.08
Lignite	0.00	0.00	0.00	0.00	0.00
Oil	-0.02	0.32	0.42	0.44	0.19
Gas	1.12	8.41	3.56	2.39	1.40
Electricity	1.17	1.45	0.97	0.91	1.14
Basic Materials	0.03	0.09	0.14	0.24	0.25
Equipment	-0.08	0.05	0.14	0.17	0.10
Consumption Goods	-0.01	-0.04	-0.07	-0.08	0.00
Construction	-0.21	-0.10	0.09	-0.12	-0.13
Transport	-0.02	-0.03	-0.02	-0.02	0.04
Private Services	-0.03	-0.13	-0.23	-0.35	-0.19
Public Services	0.03	-0.08	-0.26	-0.44	-0.39
Macroeconomy					
CO_2	0.65	-0.25	0.04	0.07	0.41
Hard Coal	2.16	-1.13	-0.31	0.04	1.48
Lignite	0.00	-0.06	-0.14	-0.29	-0.11
Oil	-0.04	-0.03	-0.12	-0.21	-0.16
Gas	0.07	2.14	2.74	2.04	1.28
GDP	-0.01	-0.14	-0.22	-0.29	-0.13
Consumption	-0.11	-0.57	-0.87	-1.29	-0.81
Investment	-0.53	-0.30	0.15	-0.30	-0.46
Export	0.00	0.09	0.10	0.30	0.21
Import	-0.20	-0.01	0.07	-0.05	-0.15
Employment	-0.07	0.04	0.08	-0.01	-0.02
Capital Stock	0.00	0.03	0.02	0.03	0.03
Exchange Rate	0.02	0.03	0.03	0.03	-0.02

Table B7. Effects of Nuclear Phase Out in the EU

	2000	2005	2010	2015	2020
Electricity Industry					
Total Generation	-0.03	0.02	0.01	0.02	0.02
HC	5.60	3.41	3.65	3.64	5.23
BC	0.01	-0.04	0.00	0.08	20.29
Oil	12.98	5.50	0.90	1.36	1.69
GAS	9.73	46.93	33.33	27.04	14.75
NUC	-7.69	-18.52	-17.71	-17.14	-16.83
Others	0.53	0.13	-0.12	0.14	-0.09
Imports	-7.55	12.93	13.03	14.61	19.79
Exports	-7.70	13.26	13.33	14.87	20.16
CO_2	5.70	7.77	7.54	7.25	9.30
Sectoral Output					
Agriculture	0.00	-0.04	-0.04	-0.05	-0.03
Hard Coal	0.83	-0.92	-1.10	-0.72	0.78
Lignite	-0.14	-0.31	-0.30	-0.30	6.75
Oil	0.15	0.33	0.20	0.09	0.10
Gas	0.06	2.87	2.79	2.62	1.32
Electricity	-0.03	0.02	0.01	0.02	0.02
Basic Materials	-0.06	-0.19	-0.17	-0.14	-0.10
Equipment	0.06	0.21	0.11	0.09	0.01
Consumption Goods	0.01	-0.04	-0.03	-0.03	0.01
Construction	-0.10	-0.08	-0.10	-0.10	-0.09
Transport	-0.04	-0.08	-0.07	-0.08	-0.05
Private Services	-0.01	-0.08	-0.09	-0.10	-0.06
Public Services	-0.01	-0.02	-0.03	-0.04	-0.02
Macroeconomy					
CO_2	1.28	0.71	0.86	0.87	1.39
Hard Coal	3.45	2.73	3.27	3.14	2.93
Lignite	-0.15	-0.44	-0.44	-0.44	9.45
Oil	0.75	0.37	0.07	0.11	0.17
Gas	0.59	-0.18	0.43	0.55	1.09
GDP	-0.02	-0.17	-0.19	-0.18	-0.09
Consumption	-0.05	-0.29	-0.25	-0.25	-0.14
Investment	-0.19	-0.15	-0.27	-0.28	-0.35
Export	-0.05	0.04	0.04	0.01	0.01
Import	-0.07	0.06	0.05	0.01	-0.06
Employment	-0.03	0.01	0.02	0.00	0.00
Capital Stock	0.01	0.02	-0.02	-0.03	-0.03
Exchange Rate	0.02	0.03	0.03	0.03	-0.02

Appendix C: European Carbon/Energy Tax - Detailed Results

Table C1. Effects of European Carbon/Energy Tax in Germany

	2000	2005	2010	2015	2020
Electricity Industry					
Total Generation	-5.32	-7.36	-9.91	-8.77	0.60
HC	-14.42	-26.66	-54.03	-69.09	-38.82
BC	-3.11	-0.16	-25.39	-27.94	-11.22
Oil	0.00	-64.21	2891.67	3411.11	-97.15
GAS	0.00	11.70	208.09	217.24	90.76
NUC	0.00	0.00	0.00	0.00	0.00
Others	0.00	0.00	0.00	0.00	0.00
Imports	31.39	64.30	88.24	69.78	-16.03
Exports	-36.85	-38.73	-33.82	-20.96	-25.64
CO_2	-8.11	-11.45	-30.75	-34.43	-12.82
Sectoral Output					
Agriculture	-0.03	-0.15	-0.36	-0.41	-0.47
Hard Coal	0.00	0.00	0.00	0.00	0.00
Lignite	-2.43	-1.23	-18.00	-19.07	-7.98
Oil	-0.25	-1.10	-0.66	-0.42	-3.84
Gas	0.24	0.28	8.04	11.53	8.96
Electricity	-5.32	-7.36	-9.91	-8.77	0.60
Basic Materials	-0.14	-0.29	-0.74	-0.61	-0.70
Equipment	0.21	0.16	0.41	0.44	0.06
Consumption Goods	0.19	0.11	0.18	0.14	0.07
Construction	0.21	-0.43	-0.55	-0.64	-0.77
Transport	0.03	-0.09	-0.13	-0.12	-0.15
Private Services	0.12	-0.12	-0.25	-0.40	-0.42
Public Services	-0.21	-0.38	-0.84	-0.95	-0.88
Macroeconomy					
CO_2	-3.02	-5.13	-12.51	-13.51	-7.23
Hard Coal	-9.43	-17.39	-34.74	-41.60	-24.66
Lignite	-2.47	-0.60	-19.67	-20.90	-7.51
Oil	-0.23	-1.08	-1.66	-1.62	-3.24
Gas	-0.49	-1.21	-2.63	-1.06	0.22
GDP	-0.01	-0.27	-0.80	-0.94	-0.74
Consumption	0.39	-0.17	-0.38	-0.88	-0.75
Investment	0.17	-1.05	-1.61	-1.93	-2.19
Export	-0.30	-0.19	-0.48	-0.33	-0.57
Import	0.06	-0.23	-0.05	-0.07	-0.80
Employment	0.18	-0.03	0.06	0.06	0.00
Capital Stock	0.02	-0.01	-0.05	-0.24	-0.38
Exchange Rate	-0.07	-0.24	-0.56	-0.73	-0.91

Table C2. Effects of European Carbon/Energy Tax in France

	2000	2005	2010	2015	2020
Electricity Industry					
Total Generation	0.80	0.41	2.45	0.56	2.30
HC	6.79	-10.05	-36.30	-40.91	-33.44
BC	0.00	0.00	0.00	0.00	0.00
Oil	0.00	1717.18	299.74	0.00	-89.72
GAS	8.33	69.82	1153.25	292.99	478.01
NUC	0.00	0.00	0.00	-0.01	0.00
Others	-0.14	2.49	0.11	1.78	3.29
Imports	-6.95	-31.29	-35.39	32.88	104.09
Exports	2.82	-2.63	8.19	4.86	21.82
CO_2	7.42	-1.62	-11.99	-23.27	-14.88
Sectoral Output					
Agriculture	-0.03	-0.11	-0.33	-0.27	-0.32
Hard Coal	-5.01	-13.69	-30.50	-31.16	-28.28
Lignite	0.00	0.00	0.00	0.00	0.00
Oil	-0.60	-0.50	-2.77	-2.59	-2.82
Gas	-0.45	-0.79	0.45	0.48	0.84
Electricity	0.80	0.41	2.45	0.56	2.30
Basic Materials	-0.28	-0.36	-1.14	-0.67	-0.78
Equipment	-0.22	-0.21	-0.76	-0.23	-0.21
Consumption Goods	-0.04	-0.08	-0.32	-0.01	-0.05
Construction	-0.23	-0.43	-0.80	-0.61	-0.36
Transport	-0.09	-0.16	-0.44	-0.13	-0.12
Private Services	0.00	-0.11	-0.21	-0.03	-0.03
Public Services	-0.12	-0.27	-0.67	-0.45	-0.41
Macroeconomy					
CO_2	-0.17	-2.26	-5.88	-6.96	-5.84
Hard Coal	2.80	-8.62	-25.59	-29.79	-25.55
Lignite	0.00	0.00	0.00	0.00	0.00
Oil	-0.64	0.09	-2.92	-2.94	-3.42
Gas	-1.70	-2.89	3.54	2.06	3.30
GDP	-0.07	-0.17	-0.37	-0.16	-0.13
Consumption	0.09	-0.28	-0.16	-0.35	-0.31
Investment	-0.44	-1.00	-1.63	-1.91	-1.37
Export	-0.30	-0.37	-1.09	-0.81	-0.73
Import	-0.21	-0.58	-1.13	-1.16	-0.95
Employment	-0.08	-0.17	-0.39	-0.03	0.06
Capital Stock	0.01	-0.01	-0.05	-0.17	-0.26
Exchange Rate	-0.07	-0.24	-0.56	-0.73	-0.91

Table C3. Effects of European Carbon/Energy Tax in U.K.

	2000	2005	2010	2015	2020
Electricity Industry					
Total Generation	-0.57	-1.59	-1.22	0.73	2.02
HC	31.08	1.31	-16.30	-18.01	-13.59
BC	0.00	0.00	0.00	0.00	0.00
Oil	0.00	0.00	-8.66	-36.56	0.00
GAS	-14.59	-5.00	16.18	38.38	25.24
NUC	0.00	0.00	0.00	0.00	0.00
Others	0.00	0.00	0.00	0.00	0.00
Imports	107.44	0.00	100.12	-36.26	-55.74
Exports	-7.52	-30.24	-32.67	23.99	143.60
CO_2	8.25	-0.99	-8.39	-9.91	-4.68
Sectoral Output					
Agriculture	0.13	0.10	0.10	-0.63	-0.63
Hard Coal	12.74	0.51	-11.59	-15.35	-12.57
Lignite	0.00	0.00	0.00	0.00	0.00
Oil	-0.20	-0.52	-1.24	-1.80	-1.82
Gas	-1.95	-2.23	-3.39	-3.14	-3.72
Electricity	-0.57	-1.59	-1.22	0.73	2.02
Basic Materials	0.41	0.62	1.07	-0.03	0.01
Equipment	0.34	0.54	0.70	-0.54	-0.42
Consumption Goods	0.23	0.35	0.59	-0.33	-0.27
Construction	-1.10	-1.05	-2.51	-1.66	-2.07
Transport	0.09	0.16	0.20	-0.43	-0.41
Private Services	-0.01	0.02	0.03	-0.41	-0.49
Public Services	0.01	-0.10	-0.16	-0.66	-0.56
Macroeconomy					
CO_2	0.56	-2.70	-7.28	-7.80	-6.72
Hard Coal	16.86	-1.47	-16.66	-18.16	-14.67
Lignite	0.00	0.00	0.00	0.00	0.00
Oil	-1.11	-2.34	-6.34	-9.23	-6.17
Gas	-4.05	-3.49	-3.65	-2.43	-4.00
GDP	-0.02	0.10	0.21	-0.36	-0.42
Consumption	-0.67	-0.81	-1.38	-0.81	-1.32
Investment	-2.37	-2.60	-5.65	-3.46	-4.23
Export	0.26	0.28	0.40	-0.44	-0.35
Import	-0.43	-0.78	-1.78	-1.51	-1.72
Employment	0.08	0.19	0.27	-0.35	-0.28
Capital Stock	-0.05	-0.24	-0.60	-0.99	-1.22
Exchange Rate	-0.07	-0.24	-0.56	-0.73	-0.91

Table C4. Effects of European Carbon/Energy Tax in Italy

	2000	2005	2010	2015	2020
Electricity Industry					
Total Generation	6.50	9.60	9.36	6.33	-8.04
HC	-3.71	-10.10	-21.21	-31.36	-40.58
BC	0.00	-22.56	-46.95	-73.17	-31.50
Oil	27.34	-3.77	2.22	-0.03	-57.60
GAS	0.06	209.24	50.47	33.82	29.54
NUC	0.00	0.00	0.00	0.00	0.00
Others	-0.89	1.02	6.95	7.17	3.29
Imports	-15.90	-24.24	-29.89	-28.81	54.53
Exports	0.00	0.00	0.00	-89.81	-97.47
CO_2	7.88	0.73	-0.60	-5.46	-22.24
Sectoral Output					
Agriculture	0.04	-0.12	-0.24	-0.14	0.09
Hard Coal	-22.79	-42.25	-60.72	-58.98	-58.21
Lignite	0.00	0.00	0.00	0.00	0.00
Oil	0.10	-1.65	-2.85	-2.40	-3.17
Gas	-0.22	11.06	12.21	11.95	10.61
Electricity	6.50	9.60	9.36	6.33	-8.04
Basic Materials	0.02	-0.55	-1.00	-0.85	-0.56
Equipment	-0.20	-0.09	-0.28	0.07	0.32
Consumption Goods	0.04	-0.14	-0.19	0.01	0.35
Construction	-0.36	-0.02	-0.37	0.09	0.21
Transport	-0.04	-0.16	-0.33	-0.15	0.08
Private Services	0.03	-0.09	-0.22	-0.18	-0.01
Public Services	-0.03	-0.30	-0.84	-1.06	-0.80
Macroeconomy					
CO_2	-0.26	-5.99	-7.17	-7.22	-11.23
Hard Coal	-5.16	-9.52	-17.88	-22.85	-28.31
Lignite	-0.01	-24.65	-48.26	-73.71	-31.34
Oil	3.39	-2.21	-2.76	-2.47	-7.65
Gas	-5.39	-12.87	-8.32	-4.09	-4.64
GDP	0.05	-0.31	-0.68	-0.64	-0.46
Consumption	0.08	-0.02	-0.29	-0.60	-0.74
Investment	-1.04	0.27	-0.61	0.02	-0.23
Export	-0.16	-0.74	-1.19	-1.03	-0.81
Import	-0.39	-0.07	-0.12	0.06	0.17
Employment	-0.11	0.02	-0.08	0.10	0.04
Capital Stock	0.01	0.05	0.15	0.21	0.31
Exchange Rate	-0.07	-0.24	-0.56	-0.73	-0.91

Table C5. Effects of European Carbon/Energy Tax in Belgium/Netherlands

	2000	2005	2010	2015	2020
Electricity Industry					
Total Generation	10.92	14.41	12.37	19.75	8.50
HC	0.00	-8.30	-40.51	-47.79	-42.63
BC	0.00	0.00	0.00	0.00	0.00
Oil	0.00	0.00	16.17	0.00	-66.39
GAS	78.44	58.59	68.51	79.19	33.62
NUC	0.00	0.00	0.00	0.00	0.00
Others	0.00	0.00	0.00	0.00	0.00
Imports	-34.53	-41.83	-6.75	-26.41	-58.45
Exports	49.83	199.30	150.30	186.74	33.97
CO_2	9.80	7.74	-6.37	-1.04	-5.30
Sectoral Output					
Agriculture	-0.05	-0.34	-0.74	-0.85	-0.52
Hard Coal	-26.30	-44.19	-68.24	-66.76	-60.90
Lignite	0.00	0.00	0.00	0.00	0.00
Oil	-0.79	-1.55	-3.01	-2.60	-2.49
Gas	-0.03	0.15	0.38	0.77	-0.02
Electricity	10.92	14.41	12.37	19.75	8.50
Basic Materials	-0.28	-1.21	-2.70	-3.12	-2.03
Equipment	-0.17	-0.58	-1.53	-2.22	-1.01
Consumption Goods	0.01	-0.40	-0.86	-1.07	-0.29
Construction	-0.16	0.42	0.18	-0.29	-0.22
Transport	0.00	-0.29	-0.67	-0.88	-0.30
Private Services	0.11	0.13	0.15	0.14	0.12
Public Services	-0.17	-0.33	-0.69	-0.36	-0.33
Macroeconomy					
CO_2	-0.70	-2.56	-7.97	-7.39	-7.31
Hard Coal	-1.54	-8.10	-29.66	-32.82	-28.23
Lignite	0.00	0.00	0.00	0.00	0.00
Oil	-0.57	-0.91	-1.93	-1.76	-2.20
Gas	-0.31	-1.02	-2.30	-0.92	-3.65
GDP	0.05	-0.02	-0.36	-0.37	-0.32
Consumption	0.33	0.98	1.91	2.29	0.90
Investment	-0.61	1.20	0.79	-0.07	-0.86
Export	-0.23	-0.39	-0.84	-0.60	-0.67
Import	-0.31	-0.30	-0.77	-0.86	-0.85
Employment	0.06	0.11	-0.13	-0.24	0.09
Capital Stock	0.06	0.11	0.31	0.28	0.20
Exchange Rate	-0.07	-0.24	-0.56	-0.73	-0.91

Table C6. Effects of European Carbon/Energy Tax in Spain/Portugal

	2000	2005	2010	2015	2020
Electricity Industry					
Total Generation	-3.34	-2.47	-4.69	-7.00	-12.06
HC	-10.07	-14.31	-16.56	-40.77	-42.01
BC	0.00	0.00	-70.17	0.00	-38.19
Oil	0.00	0.00	0.00	0.00	0.00
GAS	0.00	16.53	10.52	7.97	-4.54
NUC	0.00	0.00	0.00	0.00	0.00
Others	-0.69	4.21	1.61	0.85	-2.63
Imports	50.42	47.55	134.51	221.29	384.54
Exports	0.00	-48.98	0.00	-63.51	-44.26
CO_2	-7.25	-8.40	-17.42	-21.10	-23.81
Sectoral Output					
Agriculture	-0.06	-0.07	-0.10	0.17	0.69
Hard Coal	-4.91	-8.41	-13.50	-18.98	-17.25
Lignite	0.00	-0.18	-61.99	-1.77	-32.19
Oil	-0.66	-1.11	-1.86	-1.40	-3.50
Gas	0.41	8.54	7.94	6.62	-2.00
Electricity	-3.34	-2.47	-4.69	-7.00	-12.06
Basic Materials	-0.08	-0.11	-0.12	0.21	-0.29
Equipment	-0.06	0.16	0.69	1.31	0.38
Consumption Goods	0.02	0.08	0.12	0.43	1.22
Construction	-0.08	0.32	0.41	0.83	2.24
Transport	-0.19	-0.28	-0.57	-0.34	0.06
Private Services	-0.06	-0.13	-0.80	-0.81	1.44
Public Services	-0.12	-0.54	-1.56	-2.58	-0.38
Macroeconomy					
CO_2	-3.25	-4.37	-8.66	-9.70	-8.97
Hard Coal	-8.07	-12.02	-15.37	-30.09	-25.51
Lignite	-0.06	-0.17	-51.81	-2.74	-26.00
Oil	-0.45	-1.19	-3.37	-2.98	-1.62
Gas	-9.30	-0.15	5.65	5.47	-1.84
GDP	-0.12	-0.28	-0.95	-0.93	1.02
Consumption	0.01	-0.29	-2.56	-3.21	3.44
Investment	-0.33	0.47	0.17	0.95	2.81
Export	-0.25	-0.40	0.10	0.52	-2.10
Import	0.00	0.08	0.23	0.38	0.29
Employment	-0.13	-0.05	0.04	0.01	-0.01
Capital Stock	0.15	0.36	0.83	1.09	1.23
Exchange Rate	-0.07	-0.24	-0.56	-0.73	-0.91

Table C7. Effects of European Carbon/Energy Tax in the EU

	2000	2005	2010	2015	2020
Electricity Industry					
Total Generation	-0.02	-0.03	0.00	0.02	0.05
HC	-0.87	-12.74	-31.30	-40.55	-31.12
BC	-2.81	-0.17	-29.12	-26.49	-14.92
Oil	13.32	3.94	7.60	-8.98	-33.79
GAS	-2.98	22.75	57.77	61.38	42.47
NUC	0.00	0.00	0.31	0.00	0.00
Others	1.12	0.86	2.70	1.40	1.13
Imports	-6.50	-8.07	3.13	12.58	18.10
Exports	-6.63	-8.27	3.20	12.81	18.44
CO_2	0.22	-4.70	-16.45	-19.30	-12.78
Sectoral Output					
Agriculture	0.00	-0.11	-0.27	-0.33	-0.21
Hard Coal	-0.75	-7.70	-16.09	-20.01	-21.23
Lignite	-1.81	-0.97	-27.37	-16.12	-13.67
Oil	-0.31	-0.89	-1.78	-1.83	-2.56
Gas	-0.62	0.37	2.34	3.32	1.96
Electricity	-0.02	-0.03	0.00	0.02	0.05
Basic Materials	-0.06	-0.28	-0.70	-0.73	-0.66
Equipment	0.05	0.07	0.05	0.01	-0.06
Consumption Goods	0.09	0.02	-0.01	-0.07	0.13
Construction	-0.31	-0.34	-0.81	-0.54	-0.46
Transport	-0.02	-0.11	-0.25	-0.29	-0.15
Private Services	0.04	-0.07	-0.19	-0.27	-0.09
Public Services	-0.10	-0.27	-0.65	-0.83	-0.64
Macroeconomy					
CO_2	-1.19	-3.95	-8.70	-9.27	-7.64
Hard Coal	-2.08	-10.60	-24.62	-30.05	-23.35
Lignite	-2.15	-0.58	-23.29	-19.27	-9.91
Oil	0.16	-1.29	-3.15	-3.63	-4.35
Gas	-2.63	-3.55	-2.63	-1.31	-2.33
GDP	-0.02	-0.17	-0.47	-0.57	-0.34
Consumption	0.03	-0.21	-0.48	-0.60	-0.34
Investment	-0.78	-0.89	-2.02	-1.61	-1.72
Export	-0.16	-0.26	-0.54	-0.51	-0.70
Import	-0.22	-0.38	-0.71	-0.64	-0.82
Employment	0.02	0.00	-0.04	-0.07	-0.03
Capital Stock	0.02	-0.01	-0.03	-0.16	-0.25
Exchange Rate	-0.07	-0.24	-0.56	-0.73	-0.91

Appendix D: National Carbon/Energy Tax in Germany - Detailed Results

Table D1. Effects of German Carbon/Energy Tax in Germany

	2000	2005	2010	2015	2020
Electricity Industry					
Total Generation	-15.40	-28.51	-55.22	-58.25	-57.96
HC	-31.22	-63.69	-90.20	-91.64	-82.12
BC	-19.64	-29.16	-89.74	-90.37	-91.52
Oil	0.00	-64.21	0.00	0.00	-97.15
GAS	0.00	-20.88	-27.10	-69.12	-86.84
NUC	0.00	0.00	0.00	0.00	0.00
Others	0.00	0.00	0.00	0.00	0.00
Imports	138.93	338.84	604.13	524.16	406.54
Exports	-62.96	-51.28	-44.83	-37.17	-34.75
CO_2	-25.16	-45.52	-88.10	-90.22	-88.52
Sectoral Output					
Agriculture	0.00	-0.17	-0.40	-0.69	-0.83
Hard Coal	0.00	0.00	0.00	0.00	0.00
Lignite	-13.23	-19.39	-52.01	-50.98	-45.77
Oil	-0.11	-0.78	-1.63	-1.74	-3.36
Gas	0.24	-0.76	-1.68	-4.49	-9.09
Electricity	-15.40	-28.51	-55.22	-58.25	-57.96
Basic Materials	0.09	-0.07	-0.45	-0.89	-0.81
Equipment	0.70	0.77	1.47	0.70	0.68
Consumption Goods	0.33	0.28	0.60	0.27	0.25
Construction	0.56	-0.41	-0.29	-1.56	-1.59
Transport	0.11	-0.11	-0.17	-0.54	-0.47
Private Services	0.07	-0.31	-0.46	-0.80	-0.91
Public Services	-0.37	-0.52	-0.94	-0.83	-0.64
Macroeconomy					
CO_2	-7.21	-13.15	-23.22	-22.44	-18.01
Hard Coal	-18.63	-37.62	-54.66	-54.20	-43.65
Lignite	-15.32	-22.52	-60.63	-59.23	-53.77
Oil	-0.14	-0.82	-1.86	-1.94	-2.59
Gas	-0.44	-0.92	-2.09	-2.37	-2.96
GDP	-0.07	-0.48	-0.82	-1.14	-0.97
Consumption	-0.03	-0.85	-1.02	-1.10	-1.41
Investment	0.56	-1.16	-1.55	-3.18	-3.42
Export	0.09	0.60	0.89	0.72	0.86
Import	0.44	0.33	0.59	0.12	-0.31
Employment	0.36	0.02	0.30	-0.18	-0.06
Capital Stock	0.01	-0.03	-0.15	-0.44	-0.69
Exchange Rate	-0.04	-0.15	-0.35	-0.48	-0.75

Table D2. Effects of German Carbon/Energy Tax in France

	2000	2005	2010	2015	2020
Electricity Industry					
Total Generation	1.60	4.12	12.75	12.34	13.19
HC	14.75	22.59	9.12	6.65	0.82
BC	0.00	0.00	0.00	0.00	0.00
Oil	0.00	2048.72	2087.18	2123.08	132.41
GAS	0.00	73.68	1839.32	532.05	860.71
NUC	0.00	0.00	0.00	0.00	0.00
Others	-0.16	-0.30	-0.44	0.36	-0.57
Imports	10.22	3.50	66.38	274.23	330.65
Exports	8.37	18.68	70.28	80.31	106.56
CO_2	14.78	34.51	59.09	52.00	52.05
Sectoral Output					
Agriculture	0.02	0.04	0.03	0.09	0.07
Hard Coal	4.30	8.17	6.55	5.13	2.40
Lignite	0.00	0.00	0.00	0.00	0.00
Oil	-0.02	0.85	0.62	0.70	0.10
Gas	0.00	0.00	3.48	3.77	4.24
Electricity	1.60	4.12	12.75	12.34	13.19
Basic Materials	0.02	0.29	0.21	0.35	0.15
Equipment	-0.14	-0.11	-0.52	-0.40	-0.61
Consumption Goods	0.01	0.01	-0.12	0.01	-0.16
Construction	-0.08	-0.05	0.04	0.29	0.17
Transport	0.02	0.07	0.06	0.23	0.10
Private Services	0.03	0.01	0.09	0.26	0.04
Public Services	0.06	0.06	0.08	0.12	0.01
Macroeconomy					
CO_2	1.72	3.74	5.55	5.24	4.52
Hard Coal	8.58	13.59	7.29	4.94	0.40
Lignite	0.00	0.00	0.00	0.00	0.00
Oil	-0.03	1.62	1.30	1.36	0.48
Gas	0.08	0.46	15.30	15.14	17.44
GDP	0.03	0.05	0.19	0.32	0.16
Consumption	0.02	-0.28	-0.03	0.15	-0.35
Investment	-0.17	-0.30	-0.11	-0.03	-0.19
Export	-0.09	0.06	-0.19	-0.04	0.10
Import	0.00	0.03	0.14	0.33	0.08
Employment	0.00	0.00	0.03	0.16	-0.13
Capital Stock	0.01	0.01	0.02	0.06	0.10
Exchange Rate	-0.04	-0.15	-0.35	-0.48	-0.75

Table D3. Effects of German Carbon/Energy Tax in U.K.

	2000	2005	2010	2015	2020
Electricity Industry					
Total Generation	0.18	0.01	0.88	3.24	3.04
HC	35.14	7.43	0.01	0.13	0.00
BC	0.00	0.00	0.00	0.00	0.00
Oil	0.00	0.00	22.64	0.00	0.00
GAS	-14.59	-5.00	-0.18	11.21	10.57
NUC	0.00	0.00	0.00	0.00	0.00
Others	0.00	0.00	0.00	0.00	0.00
Imports	56.37	0.00	24.81	-60.60	-55.74
Exports	10.39	0.23	39.19	216.80	245.26
CO_2	10.13	2.56	1.59	2.54	2.30
Sectoral Output					
Agriculture	-0.04	-0.01	-0.09	0.00	0.03
Hard Coal	14.83	5.07	0.68	0.38	0.54
Lignite	0.00	0.00	0.00	0.00	0.00
Oil	-0.01	0.06	0.18	0.17	0.12
Gas	-1.13	-0.33	-0.01	0.66	0.30
Electricity	0.18	0.01	0.88	3.24	3.04
Basic Materials	0.01	0.11	-0.03	0.00	0.13
Equipment	-0.20	-0.15	-0.70	-0.45	-0.37
Consumption Goods	-0.11	-0.02	-0.19	-0.05	-0.05
Construction	-0.55	0.04	-0.23	0.55	0.16
Transport	-0.09	0.03	-0.08	0.12	0.15
Private Services	-0.12	0.02	0.00	0.25	0.08
Public Services	-0.05	0.04	0.10	0.20	0.11
Macroeconomy					
CO_2	2.14	0.56	0.25	0.58	0.38
Hard Coal	20.81	5.28	-0.01	0.10	-0.06
Lignite	0.00	0.00	0.00	0.00	0.00
Oil	-0.11	-0.01	0.78	-0.04	-0.13
Gas	-2.82	-0.84	-0.02	1.25	0.88
GDP	-0.19	-0.01	-0.06	0.17	0.08
Consumption	-0.33	-0.10	0.01	0.48	-0.35
Investment	-0.95	-0.07	-0.61	0.73	-0.24
Export	-0.05	-0.16	-0.53	-0.66	-0.46
Import	-0.13	-0.09	-0.32	0.10	-0.31
Employment	-0.16	0.03	-0.12	0.14	0.00
Capital Stock	0.01	0.01	0.04	0.09	0.17
Exchange Rate	-0.04	-0.15	-0.35	-0.48	-0.75

Table D4. Effects of German Carbon/Energy Tax in Italy

	2000	2005	2010	2015	2020
Electricity Industry					
Total Generation	12.07	23.87	34.50	28.62	18.09
HC	6.11	2.25	4.03	0.00	12.76
BC	10.24	0.00	0.00	0.00	0.00
Oil	33.86	34.57	115.56	126.10	18.28
GAS	0.06	180.39	43.47	33.82	29.54
NUC	0.00	0.00	0.00	0.00	0.00
Others	-1.02	-2.53	3.68	1.59	-2.03
Imports	-29.50	-55.05	-83.76	-82.54	-73.32
Exports	0.00	7950.88	34410.00	4488.06	1027.40
CO_2	16.64	22.23	41.16	31.83	18.69
Sectoral Output					
Agriculture	0.05	-0.03	-0.09	-0.26	-0.30
Hard Coal	3.62	-5.86	-9.17	-15.11	-4.92
Lignite	0.00	0.00	0.00	0.00	0.00
Oil	0.69	0.64	1.92	1.98	0.39
Gas	-0.19	8.91	8.69	9.28	8.01
Electricity	12.07	23.87	34.50	28.62	18.09
Basic Materials	0.15	-0.25	-0.32	-0.86	-0.79
Equipment	-0.23	-0.14	-0.48	-0.40	-0.43
Consumption Goods	0.02	-0.15	-0.23	-0.37	-0.30
Construction	-0.31	-0.06	-0.50	-0.08	0.02
Transport	0.00	-0.03	-0.12	-0.20	-0.19
Private Services	0.02	-0.03	-0.12	-0.18	-0.25
Public Services	0.10	0.10	0.13	-0.23	-0.34
Macroeconomy					
CO_2	3.58	1.98	6.49	5.27	2.74
Hard Coal	4.19	2.96	5.79	5.90	10.73
Lignite	11.37	-0.01	-0.01	-0.01	-0.01
Oil	4.69	4.87	11.21	8.76	1.92
Gas	-0.16	-6.94	-5.56	-4.05	-2.32
GDP	0.10	-0.05	-0.13	-0.33	-0.34
Consumption	-0.04	-0.08	-0.36	-0.09	-0.48
Investment	-0.96	0.14	-1.19	0.21	0.26
Export	-0.04	-0.29	0.14	0.48	0.40
Import	-0.51	-0.19	-0.08	0.86	0.49
Employment	-0.08	0.11	0.06	0.20	-0.02
Capital Stock	-0.02	-0.02	-0.01	0.03	-0.03
Exchange Rate	-0.04	-0.15	-0.35	-0.48	-0.75

Table D5. Effects of German Carbon/Energy Tax in Belgium/Netherlands

	2000	2005	2010	2015	2020
Electricity Industry					
Total Generation	24.32	32.82	36.09	37.80	41.64
HC	2.89	4.78	4.32	15.45	119.03
BC	0.00	0.00	0.00	0.00	0.00
Oil	0.00	0.00	35.11	0.00	0.00
GAS	165.39	104.05	93.52	79.19	33.62
NUC	0.00	0.00	0.00	0.00	0.00
Others	0.00	0.00	0.00	0.00	0.00
Imports	-40.23	6.91	38.03	45.36	33.64
Exports	396.80	748.56	500.13	450.29	310.82
CO_2	20.79	29.74	33.21	37.65	65.23
Sectoral Output					
Agriculture	-0.11	-0.43	-0.58	-0.72	-0.81
Hard Coal	-1.49	-5.39	-7.16	-5.54	9.60
Lignite	0.00	0.00	0.00	0.00	0.00
Oil	-0.31	-0.17	0.07	-0.16	-0.56
Gas	0.52	0.99	1.21	1.14	0.27
Electricity	24.32	32.82	36.09	37.80	41.64
Basic Materials	-0.52	-1.41	-1.80	-1.96	-2.06
Equipment	-1.04	-1.55	-2.62	-3.02	-2.87
Consumption Goods	-0.37	-0.84	-1.16	-1.39	-1.45
Construction	-0.46	0.58	-0.36	-0.55	0.08
Transport	-0.28	-0.59	-0.86	-1.09	-1.08
Private Services	0.14	0.39	0.35	0.19	0.02
Public Services	0.32	0.52	0.80	0.78	0.41
Macroeconomy					
CO_2	2.70	4.17	4.55	4.80	10.72
Hard Coal	1.90	3.23	2.97	7.31	48.79
Lignite	0.00	0.00	0.00	0.00	0.00
Oil	-0.11	0.50	0.85	0.36	0.10
Gas	5.34	7.25	7.70	6.52	4.77
GDP	0.25	0.48	0.38	0.18	0.08
Consumption	0.97	2.22	2.60	2.56	2.04
Investment	-0.67	2.13	0.19	-0.18	1.48
Export	0.11	0.28	0.17	-0.01	0.02
Import	-0.25	0.14	-0.21	-0.30	-0.18
Employment	-0.16	0.22	-0.05	-0.24	-0.07
Capital Stock	0.01	-0.02	0.00	-0.05	-0.16
Exchange Rate	-0.04	-0.15	-0.35	-0.48	-0.75

Table D6. Effects of German Carbon/Energy Tax in Spain/Portugal

	2000	2005	2010	2015	2020
Electricity Industry					
Total Generation	1.81	2.11	3.27	2.41	1.22
HC	4.98	1.95	6.95	0.00	14.37
BC	0.00	0.00	0.00	0.00	0.00
Oil	0.00	0.00	0.00	0.00	0.00
GAS	0.00	12.41	6.50	7.97	-4.54
NUC	0.00	0.00	0.00	0.00	0.00
Others	-0.63	0.12	1.61	-0.74	-0.72
Imports	-26.58	-32.83	-45.27	-56.65	-20.86
Exports	26.51	173.87	1183.05	178.35	232.38
CO_2	3.91	2.56	5.71	2.21	4.24
Sectoral Output					
Agriculture	-0.04	-0.02	-0.03	0.03	0.16
Hard Coal	1.14	0.53	1.88	1.59	3.45
Lignite	0.00	0.00	0.00	-0.25	0.17
Oil	-0.11	0.25	0.90	1.75	-1.25
Gas	1.42	6.32	4.74	5.99	-2.47
Electricity	1.81	2.11	3.27	2.41	1.22
Basic Materials	-0.02	0.05	0.23	0.94	0.07
Equipment	-0.30	-0.10	0.02	0.84	-0.68
Consumption Goods	-0.06	-0.03	-0.09	-0.09	0.30
Construction	-0.37	-0.04	-0.14	-0.43	0.68
Transport	-0.09	0.01	0.01	0.17	0.31
Private Services	-0.05	-0.05	-0.35	-1.15	0.99
Public Services	0.06	0.07	-0.10	-1.11	0.60
Macroeconomy					
CO_2	1.01	0.53	1.53	0.19	1.76
Hard Coal	3.34	1.41	4.71	0.29	6.80
Lignite	-0.01	-0.02	-0.21	-0.98	0.57
Oil	-0.07	0.03	-0.07	-0.56	0.41
Gas	0.24	1.64	3.52	5.02	-1.69
GDP	-0.06	-0.06	-0.29	-0.87	0.93
Consumption	-0.10	-0.31	-1.52	-4.51	2.76
Investment	-0.87	-0.22	-0.58	-1.35	0.47
Export	-0.11	-0.04	0.40	1.75	-1.58
Import	-0.36	-0.07	0.10	0.15	-0.71
Employment	-0.19	0.05	0.19	0.30	0.03
Capital Stock	0.00	0.02	0.09	0.17	0.02
Exchange Rate	-0.04	-0.15	-0.35	-0.48	-0.75

Table D7. Effects of German Carbon/Energy Tax in the EU

	2000	2005	2010	2015	2020
Electricity Industry					
Total Generation	-0.02	0.08	0.23	0.27	0.24
HC	-0.36	-10.99	-18.72	-18.97	-3.13
BC	-17.70	-26.21	-81.97	-82.91	-80.04
Oil	17.81	25.83	82.60	56.62	8.72
GAS	4.67	26.15	48.26	43.64	25.80
NUC	0.00	0.00	0.31	0.00	0.00
Others	1.30	0.88	-1.54	-0.92	-2.97
Imports	-0.44	29.39	94.73	120.47	135.32
Exports	-0.45	30.15	96.91	122.63	137.84
CO_2	-1.94	-6.72	-15.80	-16.56	-10.68
Sectoral Output					
Agriculture	-0.01	-0.07	-0.15	-0.22	-0.24
Hard Coal	3.70	0.93	-0.44	-1.38	0.52
Lignite	-9.87	-14.64	-40.93	-42.35	-34.98
Oil	0.05	0.16	0.32	0.39	-0.43
Gas	-0.17	0.78	1.25	1.00	-0.82
Electricity	-0.02	0.08	0.23	0.27	0.24
Basic Materials	0.02	-0.13	-0.33	-0.51	-0.56
Equipment	0.09	0.13	0.11	-0.06	-0.18
Consumption Goods	0.04	-0.04	-0.05	-0.14	-0.13
Construction	-0.13	-0.06	-0.25	-0.29	-0.23
Transport	-0.03	-0.07	-0.15	-0.19	-0.18
Private Services	0.02	-0.06	-0.13	-0.23	-0.20
Public Services	-0.08	-0.09	-0.19	-0.21	-0.16
Macroeconomy					
CO_2	-0.43	-2.17	-3.98	-3.89	-2.12
Hard Coal	-0.57	-8.08	-13.91	-13.50	-4.26
Lignite	-13.30	-19.64	-53.81	-53.78	-46.74
Oil	0.80	1.07	2.02	1.28	-0.16
Gas	-0.18	0.07	2.23	2.63	2.22
GDP	-0.02	-0.10	-0.21	-0.32	-0.20
Consumption	0.00	-0.15	-0.22	-0.27	-0.22
Investment	-0.35	-0.23	-0.79	-0.78	-0.85
Export	0.01	0.16	0.19	0.24	0.11
Import	-0.05	0.06	0.06	0.19	-0.13
Employment	0.03	0.05	0.08	0.05	-0.05
Capital Stock	0.00	-0.01	-0.02	-0.07	-0.14
Exchange Rate	-0.04	-0.15	-0.35	-0.48	-0.75

Appendix E: European Carbon/Energy Tax under Alternative Tax Recycling - Detailed Results

Table E1a. Effects of European Carbon/Energy Tax under Alternative Tax Recycling in Germany (Low Wage Elasticity)

	2000	2005	2010	2015	2020
Electricity Industry					
Total Generation	-5.32	-7.36	-9.91	-8.77	0.60
HC	-14.42	-26.66	-54.03	-69.09	-38.82
BC	-3.11	-0.16	-25.39	-27.94	-11.22
Oil	0.00	-64.21	2891.67	3411.11	-97.15
GAS	0.00	11.70	208.09	217.24	90.76
NUC	0.00	0.00	0.00	0.00	0.00
Others	0.00	0.00	0.00	0.00	0.00
Imports	31.39	64.30	88.24	69.78	-16.03
Exports	-36.85	-38.73	-33.82	-20.96	-25.64
CO_2	-8.11	-11.45	-30.75	-34.43	-12.82
Sectoral Output					
Agriculture	0.28	0.70	1.32	1.26	1.22
Hard Coal	0.00	0.00	0.00	0.00	0.00
Lignite	-2.37	-0.88	-17.23	-18.39	-7.29
Oil	-0.12	-0.58	0.38	0.51	-2.95
Gas	0.41	0.81	8.96	12.48	9.93
Electricity	-5.32	-7.36	-9.91	-8.77	0.60
Basic Materials	0.20	0.77	1.30	1.24	1.11
Equipment	0.51	1.17	2.41	2.14	1.72
Consumption Goods	0.43	1.02	1.99	1.70	1.53
Construction	0.45	0.24	0.79	0.53	0.40
Transport	0.16	0.52	1.11	0.91	0.81
Private Services	0.29	0.57	1.14	0.82	0.75
Public Services	0.10	0.76	1.29	0.92	0.87
Macroeconomy					
CO_2	-2.87	-4.65	-11.60	-12.62	-6.29
Hard Coal	-9.31	-17.04	-34.10	-40.96	-23.89
Lignite	-2.43	-0.32	-19.02	-20.35	-6.95
Oil	-0.06	-0.51	-0.55	-0.57	-2.22
Gas	-0.29	-0.61	-1.56	0.03	1.37
GDP	0.23	0.61	0.92	0.57	0.72
Consumption	0.54	0.58	1.19	0.49	0.48
Investment	0.47	-0.55	-0.60	-0.86	-0.98
Export	-0.17	0.18	0.21	0.36	0.22
Import	0.16	0.16	0.76	0.65	-0.10
Employment	0.48	1.20	2.47	1.96	1.69
Capital Stock	0.15	0.23	0.40	0.40	0.42
Exchange Rate	0.35	0.50	0.80	0.98	0.98

Table E2a. Effects of European Carbon/Energy Tax under Alternative Tax Recycling in France (Low Wage Elasticity)

	2000	2005	2010	2015	2020
Electricity Industry					
Total Generation	0.80	0.41	2.45	0.56	2.30
HC	6.79	-10.05	-36.30	-40.91	-33.44
BC	0.00	0.00	0.00	0.00	0.00
Oil	0.00	1717.18	299.74	0.00	-89.72
GAS	8.33	69.82	1153.25	292.99	478.01
NUC	0.00	0.00	0.00	-0.01	0.00
Others	-0.14	2.49	0.11	1.78	3.29
Imports	-6.95	-31.29	-35.39	32.88	104.09
Exports	2.82	-2.63	8.19	4.86	21.82
CO_2	7.42	-1.62	-11.99	-23.27	-14.88
Sectoral Output					
Agriculture	0.39	0.59	0.83	0.98	0.93
Hard Coal	-2.44	-10.59	-26.98	-28.33	-26.25
Lignite	0.00	0.00	0.00	0.00	0.00
Oil	-0.18	0.22	-1.65	-1.55	-1.91
Gas	0.12	0.12	1.90	1.90	2.12
Electricity	0.80	0.41	2.45	0.56	2.30
Basic Materials	0.59	0.97	0.82	1.34	1.10
Equipment	0.72	1.17	1.17	1.72	1.59
Consumption Goods	0.66	1.04	1.39	1.68	1.46
Construction	0.35	0.41	0.62	0.69	0.79
Transport	0.51	0.77	0.96	1.20	1.02
Private Services	0.56	0.81	1.26	1.35	1.16
Public Services	0.76	1.12	1.55	1.62	1.33
Macroeconomy					
CO_2	0.30	-1.53	-4.72	-5.81	-4.78
Hard Coal	3.14	-8.11	-24.83	-29.03	-24.79
Lignite	0.00	0.00	0.00	0.00	0.00
Oil	-0.18	0.82	-1.73	-1.77	-2.38
Gas	-1.07	-1.92	5.00	3.50	4.63
GDP	0.62	0.92	1.33	1.48	1.32
Consumption	0.58	0.56	1.44	1.10	0.89
Investment	-0.01	-0.45	-0.47	-0.76	-0.22
Export	0.03	0.02	-0.74	-0.28	-0.13
Import	0.14	-0.07	-0.27	-0.36	-0.27
Employment	0.87	1.33	1.89	2.02	1.72
Capital Stock	0.19	0.26	0.42	0.46	0.50
Exchange Rate	0.35	0.50	0.80	0.98	0.98

Table E3a. Effects of European Carbon/Energy Tax under Alternative Tax Recycling in U.K. (Low Wage Elasticity)

	2000	2005	2010	2015	2020
Electricity Industry					
Total Generation	-0.57	-1.59	-1.22	0.73	2.02
HC	31.08	1.31	-16.30	-18.01	-13.59
BC	0.00	0.00	0.00	0.00	0.00
Oil	0.00	0.00	-8.66	-36.56	0.00
GAS	-14.59	-5.00	16.18	38.38	25.24
NUC	0.00	0.00	0.00	0.00	0.00
Others	0.00	0.00	0.00	0.00	0.00
Imports	107.44	0.00	100.12	-36.26	-55.74
Exports	-7.52	-30.24	-32.67	23.99	143.60
CO_2	8.25	-0.99	-8.39	-9.91	-4.68
Sectoral Output					
Agriculture	0.46	1.23	2.63	2.08	2.05
Hard Coal	12.83	1.83	-8.90	-13.10	-10.81
Lignite	0.00	0.00	0.00	0.00	0.00
Oil	0.02	-0.04	-0.27	-0.70	-0.73
Gas	-1.93	-1.25	-1.09	-1.01	-1.68
Electricity	-0.57	-1.59	-1.22	0.73	2.02
Basic Materials	0.74	1.91	4.09	3.28	3.33
Equipment	0.61	2.02	4.42	3.34	3.37
Consumption Goods	0.42	1.64	3.74	2.84	2.77
Construction	-0.98	-0.14	-0.44	0.04	-0.31
Transport	0.16	1.00	2.31	1.68	1.61
Private Services	0.03	0.93	2.28	1.75	1.58
Public Services	-0.13	1.02	2.66	1.96	1.94
Macroeconomy					
CO_2	0.63	-1.98	-5.66	-6.22	-5.13
Hard Coal	16.91	-1.07	-15.95	-17.44	-13.95
Lignite	0.00	0.00	0.00	0.00	0.00
Oil	-0.98	-1.65	-4.73	-7.65	-4.54
Gas	-4.00	-2.63	-1.61	-0.48	-2.10
GDP	0.07	1.18	2.85	2.23	2.08
Consumption	-0.86	-0.30	-0.13	0.14	-0.34
Investment	-2.29	-1.92	-4.28	-2.63	-3.12
Export	0.37	0.57	1.22	0.66	0.75
Import	-0.48	-0.34	-0.70	-0.56	-0.73
Employment	0.03	1.51	3.62	2.75	2.61
Capital Stock	0.23	0.18	0.08	-0.09	-0.17
Exchange Rate	0.35	0.50	0.80	0.98	0.98

Table E4a. Effects of European Carbon/Energy Tax under Alternative Tax Recycling in Italy (Low Wage Elasticity)

	2000	2005	2010	2015	2020
Electricity Industry					
Total Generation	6.50	9.60	9.36	6.33	-8.04
HC	-3.71	-10.10	-21.21	-31.36	-40.58
BC	0.00	-22.56	-46.95	-73.17	-31.50
Oil	27.34	-3.77	2.22	-0.03	-57.60
GAS	0.06	209.24	50.47	33.82	29.54
NUC	0.00	0.00	0.00	0.00	0.00
Others	-0.89	1.02	6.95	7.17	3.29
Imports	-15.90	-24.24	-29.89	-28.81	54.53
Exports	0.00	0.00	0.00	-89.81	-97.47
CO_2	7.88	0.73	-0.60	-5.46	-22.24
Sectoral Output					
Agriculture	0.43	0.71	1.18	1.36	1.52
Hard Coal	-22.32	-41.44	-59.86	-58.10	-57.50
Lignite	0.00	0.00	0.00	0.00	0.00
Oil	0.34	-1.06	-1.82	-1.38	-2.25
Gas	0.07	11.60	12.82	12.54	11.25
Electricity	6.50	9.60	9.36	6.33	-8.04
Basic Materials	0.60	0.59	0.88	1.17	1.38
Equipment	0.35	0.93	1.34	1.81	1.99
Consumption Goods	0.47	0.77	1.32	1.55	1.75
Construction	0.03	0.76	0.98	1.33	1.31
Transport	0.30	0.59	0.90	1.08	1.18
Private Services	0.35	0.64	1.01	1.05	1.11
Public Services	0.70	1.44	1.96	1.50	1.29
Macroeconomy					
CO_2	0.01	-5.45	-6.25	-6.26	-10.33
Hard Coal	-5.00	-9.22	-17.44	-22.42	-27.87
Lignite	-0.01	-24.65	-48.26	-73.71	-31.34
Oil	3.66	-1.61	-1.66	-1.31	-6.62
Gas	-5.02	-12.22	-7.40	-3.14	-3.65
GDP	0.49	0.64	0.90	0.94	0.95
Consumption	0.38	0.85	1.30	0.85	0.43
Investment	-0.67	0.93	0.61	1.09	0.82
Export	0.09	-0.40	-0.78	-0.33	0.01
Import	-0.19	0.39	0.68	0.80	0.82
Employment	0.61	1.64	2.52	2.47	2.00
Capital Stock	0.15	0.29	0.57	0.78	0.97
Exchange Rate	0.35	0.50	0.80	0.98	0.98

Table E5a. Effects of European Carbon/Energy Tax under Alternative Tax Recycling in Belgium/Netherlands (Low Wage Elasticity)

	2000	2005	2010	2015	2020
Electricity Industry					
Total Generation	10.92	14.41	12.37	19.75	8.50
HC	0.00	-8.30	-40.51	-47.79	-42.63
BC	0.00	0.00	0.00	0.00	0.00
Oil	0.00	0.00	16.17	0.00	-66.39
GAS	78.44	58.59	68.51	79.19	33.62
NUC	0.00	0.00	0.00	0.00	0.00
Others	0.00	0.00	0.00	0.00	0.00
Imports	-34.53	-41.83	-6.75	-26.41	-58.45
Exports	49.83	199.30	150.30	186.74	33.97
CO_2	9.80	7.74	-6.37	-1.04	-5.30
Sectoral Output					
Agriculture	0.24	0.29	0.65	0.74	1.08
Hard Coal	-26.05	-43.36	-66.14	-65.10	-59.55
Lignite	0.00	0.00	0.00	0.00	0.00
Oil	-0.86	-1.38	-2.42	-2.21	-2.26
Gas	0.10	0.48	1.13	1.63	0.90
Electricity	10.92	14.41	12.37	19.75	8.50
Basic Materials	0.01	-0.45	-0.63	-0.96	-0.01
Equipment	-0.11	-0.07	0.58	-0.16	0.82
Consumption Goods	0.08	0.07	0.76	0.52	1.10
Construction	-0.10	0.76	1.38	0.78	0.66
Transport	0.05	0.07	0.58	0.35	0.77
Private Services	0.08	0.43	1.33	1.19	0.95
Public Services	-0.25	0.14	1.12	1.23	0.84
Macroeconomy					
CO_2	-0.61	-2.29	-7.21	-6.55	-6.51
Hard Coal	-1.51	-7.94	-29.18	-32.32	-27.75
Lignite	0.00	0.00	0.00	0.00	0.00
Oil	-0.53	-0.66	-1.12	-0.96	-1.52
Gas	-0.17	-0.69	-1.44	0.08	-2.69
GDP	0.10	0.43	1.17	1.10	0.94
Consumption	0.13	0.96	2.28	2.48	0.81
Investment	-0.43	1.44	1.47	0.60	-0.14
Export	-0.28	-0.44	-0.66	-0.37	-0.48
Import	-0.38	-0.30	-0.41	-0.55	-0.64
Employment	-0.05	0.52	1.75	1.33	1.20
Capital Stock	0.28	0.47	0.93	1.13	1.20
Exchange Rate	0.35	0.50	0.80	0.98	0.98

Table E6a. Effects of European Carbon/Energy Tax under Alternative Tax Recycling in Spain/Portugal (Low Wage Elasticity)

	2000	2005	2010	2015	2020
Electricity Industry					
Total Generation	-3.34	-2.47	-4.69	-7.00	-12.06
HC	-10.07	-14.31	-16.56	-40.77	-42.01
BC	0.00	0.00	-70.17	0.00	-38.19
Oil	0.00	0.00	0.00	0.00	0.00
GAS	0.00	16.53	10.52	7.97	-4.54
NUC	0.00	0.00	0.00	0.00	0.00
Others	-0.69	4.21	1.61	0.85	-2.63
Imports	50.42	47.55	134.51	221.29	384.54
Exports	0.00	-48.98	0.00	-63.51	-44.26
CO_2	-7.25	-8.40	-17.42	-21.10	-23.81
Sectoral Output					
Agriculture	0.32	0.93	2.08	2.34	2.68
Hard Coal	-4.13	-6.08	-8.61	-15.94	-15.44
Lignite	0.00	0.00	-61.38	-1.01	-31.84
Oil	-0.48	-0.45	-0.55	-0.39	-2.65
Gas	0.81	8.90	8.27	6.96	-1.59
Electricity	-3.34	-2.47	-4.69	-7.00	-12.06
Basic Materials	0.40	1.18	2.64	2.61	1.72
Equipment	0.44	1.61	3.79	3.70	2.18
Consumption Goods	0.38	1.15	2.52	2.57	2.96
Construction	0.33	1.43	3.00	3.16	3.98
Transport	0.09	0.63	1.45	1.30	1.26
Private Services	0.27	0.87	1.58	1.46	3.20
Public Services	0.58	1.45	2.73	1.04	2.03
Macroeconomy					
CO_2	-3.00	-3.70	-7.19	-8.31	-7.82
Hard Coal	-7.88	-11.53	-14.24	-29.05	-24.55
Lignite	0.02	0.14	-50.65	-1.61	-25.39
Oil	-0.16	-0.41	-1.60	-1.26	-0.23
Gas	-8.81	0.60	6.29	6.01	-1.26
GDP	0.29	0.93	1.83	1.59	2.96
Consumption	0.40	0.86	0.42	-0.14	5.52
Investment	0.13	1.43	2.49	3.31	4.67
Export	-0.10	0.00	0.77	0.94	-1.54
Import	0.20	0.67	1.56	1.52	1.10
Employment	0.50	1.90	4.25	3.05	1.82
Capital Stock	0.32	0.72	1.61	2.30	2.73
Exchange Rate	0.35	0.50	0.80	0.98	0.98

Table E7a. Effects of European Carbon/Energy Tax under Alternative Tax Recycling in the EU (Low Wage Elasticity)

	2000	2005	2010	2015	2020
Electricity Industry					
Total Generation	-0.02	-0.03	0.00	0.02	0.05
HC	-0.87	-12.74	-31.30	-40.55	-31.12
BC	-2.81	-0.17	-29.12	-26.49	-14.92
Oil	13.32	3.94	7.60	-8.98	-33.79
GAS	-2.98	22.75	57.77	61.38	42.47
NUC	0.00	0.00	0.31	0.00	0.00
Others	1.12	0.86	2.70	1.40	1.13
Imports	-6.50	-8.07	3.13	12.58	18.10
Exports	-6.63	-8.27	3.20	12.81	18.44
CO_2	0.22	-4.70	-16.45	-19.30	-12.78
Sectoral Output					
Agriculture	0.36	0.74	1.39	1.41	1.50
Hard Coal	-0.50	-6.83	-14.29	-18.44	-19.99
Lignite	-1.77	-0.66	-26.63	-15.43	-13.06
Oil	-0.09	-0.35	-0.76	-0.81	-1.61
Gas	-0.44	1.04	3.67	4.61	3.23
Electricity	-0.02	-0.03	0.00	0.02	0.05
Basic Materials	0.41	0.86	1.50	1.47	1.45
Equipment	0.50	1.23	2.33	2.19	2.00
Consumption Goods	0.44	1.01	1.99	1.85	1.88
Construction	0.01	0.44	0.77	0.85	0.83
Transport	0.23	0.64	1.26	1.13	1.12
Private Services	0.29	0.70	1.38	1.21	1.24
Public Services	0.24	0.98	1.85	1.43	1.35
Macroeconomy					
CO_2	-0.99	-3.38	-7.55	-8.13	-6.52
Hard Coal	-1.94	-10.23	-23.94	-29.37	-22.65
Lignite	-2.11	-0.29	-22.59	-18.67	-9.34
Oil	0.40	-0.67	-1.89	-2.37	-3.19
Gas	-2.42	-2.85	-1.22	0.09	-0.95
GDP	0.32	0.80	1.46	1.26	1.33
Consumption	0.20	0.50	1.04	0.74	0.77
Investment	-0.49	-0.31	-0.81	-0.52	-0.56
Export	-0.02	0.02	-0.01	0.12	-0.01
Import	-0.12	0.00	0.11	0.09	-0.16
Employment	0.45	1.35	2.64	2.23	1.90
Capital Stock	0.20	0.29	0.50	0.57	0.63
Exchange Rate	0.35	0.50	0.80	0.98	0.98

Table E1b. Effects of European Carbon/Energy Tax under Alternative Tax Recycling in Germany (High Wage Elasticity)

	2000	2005	2010	2015	2020
Electricity Industry					
Total Generation	-5.32	-7.36	-9.91	-8.77	0.60
HC	-14.42	-26.66	-54.03	-69.09	-38.82
BC	-3.11	-0.16	-25.39	-27.94	-11.22
Oil	0.00	-64.21	2891.67	3411.11	-97.15
GAS	0.00	11.70	208.09	217.24	90.76
NUC	0.00	0.00	0.00	0.00	0.00
Others	0.00	0.00	0.00	0.00	0.00
Imports	31.39	64.30	88.24	69.78	-16.03
Exports	-36.85	-38.73	-33.82	-20.96	-25.64
CO_2	-8.11	-11.45	-30.75	-34.43	-12.82
Sectoral Output					
Agriculture	0.14	0.20	0.42	0.36	0.36
Hard Coal	0.00	0.00	0.00	0.00	0.00
Lignite	-2.31	-1.05	-17.61	-18.76	-7.61
Oil	-0.13	-0.87	-0.16	0.02	-3.38
Gas	0.35	0.50	8.46	11.95	9.42
Electricity	-5.32	-7.36	-9.91	-8.77	0.60
Basic Materials	0.08	0.15	0.19	0.28	0.24
Equipment	0.43	0.59	1.33	1.30	0.98
Consumption Goods	0.41	0.52	1.04	0.91	0.86
Construction	0.36	-0.15	0.05	-0.08	-0.18
Transport	0.18	0.18	0.46	0.39	0.38
Private Services	0.28	0.19	0.43	0.19	0.21
Public Services	0.05	0.12	0.19	-0.03	0.01
Macroeconomy					
CO_2	-2.92	-4.93	-12.08	-13.10	-6.77
Hard Coal	-9.35	-17.24	-34.45	-41.30	-24.29
Lignite	-2.40	-0.46	-19.34	-20.62	-7.18
Oil	-0.11	-0.83	-1.13	-1.14	-2.73
Gas	-0.38	-0.96	-2.14	-0.57	0.76
GDP	0.19	0.12	0.02	-0.20	0.03
Consumption	0.59	0.20	0.42	-0.24	-0.11
Investment	0.26	-0.88	-1.21	-1.49	-1.68
Export	-0.24	-0.07	-0.22	-0.01	-0.19
Import	0.16	-0.06	0.34	0.26	-0.45
Employment	0.49	0.53	1.23	1.03	0.95
Capital Stock	0.05	0.07	0.13	0.01	-0.05
Exchange Rate	0.03	0.00	0.00	-0.02	-0.10

Table E2b. Effects of European Carbon/Energy Tax under Alternative Tax Recycling in France (High Wage Elasticity)

	2000	2005	2010	2015	2020
Electricity Industry					
Total Generation	0.80	0.41	2.45	0.56	2.30
HC	6.79	-10.05	-36.30	-40.91	-33.44
BC	0.00	0.00	0.00	0.00	0.00
Oil	0.00	1717.18	299.74	0.00	-89.72
GAS	8.33	69.82	1153.25	292.99	478.01
NUC	0.00	0.00	0.00	-0.01	0.00
Others	-0.14	2.49	0.11	1.78	3.29
Imports	-6.95	-31.29	-35.39	32.88	104.09
Exports	2.82	-2.63	8.19	4.86	21.82
CO_2	7.42	-1.62	-11.99	-23.27	-14.88
Sectoral Output					
Agriculture	0.08	0.13	0.21	0.27	0.26
Hard Coal	-4.30	-12.77	-28.64	-29.94	-27.30
Lignite	0.00	0.00	0.00	0.00	0.00
Oil	-0.47	-0.25	-2.22	-2.12	-2.37
Gas	-0.29	-0.48	1.15	1.10	1.45
Electricity	0.80	0.41	2.45	0.56	2.30
Basic Materials	-0.06	0.06	-0.19	0.22	0.09
Equipment	0.00	0.22	0.22	0.64	0.64
Consumption Goods	0.15	0.29	0.53	0.74	0.67
Construction	-0.09	-0.15	-0.15	-0.04	0.19
Transport	0.08	0.15	0.26	0.46	0.43
Private Services	0.17	0.21	0.52	0.59	0.56
Public Services	0.12	0.21	0.39	0.49	0.43
Macroeconomy					
CO_2	-0.05	-2.02	-5.33	-6.46	-5.35
Hard Coal	2.89	-8.45	-25.23	-29.46	-25.20
Lignite	0.00	0.00	0.00	0.00	0.00
Oil	-0.51	0.34	-2.36	-2.43	-2.93
Gas	-1.54	-2.57	4.24	2.69	3.92
GDP	0.12	0.20	0.46	0.58	0.57
Consumption	0.26	0.05	0.60	0.28	0.28
Investment	-0.36	-0.82	-1.20	-1.45	-0.87
Export	-0.27	-0.30	-0.91	-0.61	-0.51
Import	-0.12	-0.41	-0.73	-0.83	-0.64
Employment	0.20	0.35	0.76	0.91	0.90
Capital Stock	0.04	0.06	0.12	0.07	0.04
Exchange Rate	0.03	0.00	0.00	-0.02	-0.10

Table E3b. Effects of European Carbon/Energy Tax under Alternative Tax Recycling in U.K. (High Wage Elasticity)

	2000	2005	2010	2015	2020
Electricity Industry					
Total Generation	-0.57	-1.59	-1.22	0.73	2.02
HC	31.08	1.31	-16.30	-18.01	-13.59
BC	0.00	0.00	0.00	0.00	0.00
Oil	0.00	0.00	-8.66	-36.56	0.00
GAS	-14.59	-5.00	16.18	38.38	25.24
NUC	0.00	0.00	0.00	0.00	0.00
Others	0.00	0.00	0.00	0.00	0.00
Imports	107.44	0.00	100.12	-36.26	-55.74
Exports	-7.52	-30.24	-32.67	23.99	143.60
CO_2	8.25	-0.99	-8.39	-9.91	-4.68
Sectoral Output					
Agriculture	0.34	0.55	1.15	0.44	0.42
Hard Coal	13.12	1.11	-10.52	-14.53	-11.97
Lignite	0.00	0.00	0.00	0.00	0.00
Oil	-0.14	-0.36	-0.86	-1.38	-1.40
Gas	-1.72	-1.76	-2.34	-2.22	-2.91
Electricity	-0.57	-1.59	-1.22	0.73	2.02
Basic Materials	0.63	1.12	2.25	1.20	1.20
Equipment	0.63	1.15	2.14	0.86	0.88
Consumption Goods	0.50	0.91	1.89	0.87	0.84
Construction	-0.90	-0.67	-1.63	-0.90	-1.36
Transport	0.27	0.52	1.06	0.35	0.30
Private Services	0.20	0.44	1.00	0.45	0.29
Public Services	0.30	0.47	1.13	0.45	0.39
Macroeconomy					
CO_2	0.71	-2.38	-6.56	-7.14	-6.10
Hard Coal	16.96	-1.30	-16.36	-17.87	-14.40
Lignite	0.00	0.00	0.00	0.00	0.00
Oil	-0.98	-2.05	-5.65	-8.58	-5.54
Gas	-3.85	-3.09	-2.73	-1.60	-3.25
GDP	0.22	0.59	1.33	0.66	0.51
Consumption	-0.50	-0.50	-0.68	-0.28	-0.85
Investment	-2.20	-2.31	-5.01	-2.91	-3.68
Export	0.28	0.35	0.63	-0.14	-0.06
Import	-0.32	-0.57	-1.28	-1.09	-1.34
Employment	0.40	0.83	1.71	0.87	0.77
Capital Stock	-0.01	-0.13	-0.33	-0.62	-0.78
Exchange Rate	0.03	0.00	0.00	-0.02	-0.10

Table E4b. Effects of European Carbon/Energy Tax under Alternative Tax Recycling in Italy (High Wage Elasticity)

	2000	2005	2010	2015	2020
Electricity Industry					
Total Generation	6.50	9.60	9.36	6.33	-8.04
HC	-3.71	-10.10	-21.21	-31.36	-40.58
BC	0.00	-22.56	-46.95	-73.17	-31.50
Oil	27.34	-3.77	2.22	-0.03	-57.60
GAS	0.06	209.24	50.47	33.82	29.54
NUC	0.00	0.00	0.00	0.00	0.00
Others	-0.89	1.02	6.95	7.17	3.29
Imports	-15.90	-24.24	-29.89	-28.81	54.53
Exports	0.00	0.00	0.00	-89.81	-97.47
CO_2	7.88	0.73	-0.60	-5.46	-22.24
Sectoral Output					
Agriculture	0.17	0.15	0.36	0.46	0.70
Hard Coal	-22.65	-42.00	-60.36	-58.63	-57.92
Lignite	0.00	0.00	0.00	0.00	0.00
Oil	0.20	-1.46	-2.43	-2.01	-2.78
Gas	-0.13	11.22	12.45	12.18	10.87
Electricity	6.50	9.60	9.36	6.33	-8.04
Basic Materials	0.19	-0.21	-0.24	-0.07	0.23
Equipment	-0.06	0.19	0.36	0.73	0.98
Consumption Goods	0.18	0.13	0.44	0.60	0.93
Construction	-0.24	0.20	0.12	0.56	0.68
Transport	0.07	0.06	0.16	0.30	0.52
Private Services	0.15	0.14	0.29	0.29	0.47
Public Services	0.25	0.23	0.28	-0.05	0.10
Macroeconomy					
CO_2	-0.18	-5.83	-6.81	-6.86	-10.86
Hard Coal	-5.11	-9.43	-17.71	-22.69	-28.13
Lignite	-0.01	-24.65	-48.26	-73.71	-31.34
Oil	3.48	-2.03	-2.33	-2.04	-7.22
Gas	-5.28	-12.68	-7.96	-3.74	-4.25
GDP	0.20	-0.03	-0.04	-0.03	0.13
Consumption	0.24	0.28	0.36	-0.04	-0.21
Investment	-0.96	0.43	-0.25	0.41	0.23
Export	-0.15	-0.69	-1.06	-0.82	-0.57
Import	-0.32	0.06	0.17	0.32	0.43
Employment	0.16	0.52	0.99	1.02	0.88
Capital Stock	0.03	0.11	0.29	0.42	0.56
Exchange Rate	0.03	0.00	0.00	-0.02	-0.10

Table E5b. Effects of European Carbon/Energy Tax under Alternative Tax Recycling in Belgium/Netherlands (High Wage Elasticity)

	2000	2005	2010	2015	2020
Electricity Industry					
Total Generation	10.92	14.41	12.37	19.75	8.50
HC	0.00	-8.30	-40.51	-47.79	-42.63
BC	0.00	0.00	0.00	0.00	0.00
Oil	0.00	0.00	16.17	0.00	-66.39
GAS	78.44	58.59	68.51	79.19	33.62
NUC	0.00	0.00	0.00	0.00	0.00
Others	0.00	0.00	0.00	0.00	0.00
Imports	-34.53	-41.83	-6.75	-26.41	-58.45
Exports	49.83	199.30	150.30	186.74	33.97
CO_2	9.80	7.74	-6.37	-1.04	-5.30
Sectoral Output					
Agriculture	0.08	-0.03	-0.02	-0.02	0.36
Hard Coal	-25.56	-42.95	-66.67	-65.37	-59.32
Lignite	0.00	0.00	0.00	0.00	0.00
Oil	-0.68	-1.34	-2.57	-2.23	-2.15
Gas	0.05	0.34	0.78	1.23	0.50
Electricity	10.92	14.41	12.37	19.75	8.50
Basic Materials	-0.02	-0.64	-1.42	-1.76	-0.64
Equipment	0.15	0.12	0.04	-0.61	0.63
Consumption Goods	0.24	0.09	0.27	0.07	0.85
Construction	0.01	0.78	1.00	0.49	0.56
Transport	0.18	0.09	0.20	0.00	0.59
Private Services	0.30	0.51	1.01	0.94	0.92
Public Services	0.13	0.28	0.66	0.93	0.86
Macroeconomy					
CO_2	-0.59	-2.34	-7.49	-6.88	-6.77
Hard Coal	-1.46	-7.94	-29.33	-32.48	-27.84
Lignite	0.00	0.00	0.00	0.00	0.00
Oil	-0.46	-0.67	-1.38	-1.22	-1.66
Gas	-0.19	-0.79	-1.79	-0.35	-3.05
GDP	0.27	0.45	0.70	0.68	0.73
Consumption	0.41	1.14	2.29	2.52	1.08
Investment	-0.55	1.32	1.10	0.25	-0.47
Export	-0.19	-0.30	-0.63	-0.35	-0.42
Import	-0.24	-0.15	-0.43	-0.55	-0.55
Employment	0.39	0.77	1.34	1.10	1.35
Capital Stock	0.09	0.20	0.54	0.61	0.60
Exchange Rate	0.03	0.00	0.00	-0.02	-0.10

Table E6b. Effects of European Carbon/Energy Tax under Alternative Tax Recycling in Spain/Portugal (High Wage Elasticity)

	2000	2005	2010	2015	2020
Electricity Industry					
Total Generation	-3.34	-2.47	-4.69	7.00	-12.06
HC	-10.07	-14.31	-16.56	-40.77	-42.01
BC	0.00	0.00	-70.17	0.00	-38.19
Oil	0.00	0.00	0.00	0.00	0.00
GAS	0.00	16.53	10.52	7.97	-4.54
NUC	0.00	0.00	0.00	0.00	0.00
Others	-0.69	4.21	1.61	0.85	-2.63
Imports	50.42	47.55	134.51	221.29	384.54
Exports	0.00	-48.98	0.00	-63.51	-44.26
CO_2	-7.25	-8.40	-17.42	-21.10	-23.81
Sectoral Output					
Agriculture	0.14	0.36	0.92	1.29	1.86
Hard Coal	-4.37	-7.30	-11.13	-17.06	-15.62
Lignite	0.00	0.00	-61.59	-1.26	-32.01
Oil	-0.51	-0.80	-1.23	-0.82	-2.89
Gas	0.61	8.72	8.09	6.77	-1.78
Electricity	-3.34	-2.47	-4.69	-7.00	-12.06
Basic Materials	0.17	0.44	1.15	1.52	1.01
Equipment	0.24	0.82	2.16	2.76	1.79
Consumption Goods	0.24	0.56	1.25	1.62	2.39
Construction	0.16	0.82	1.59	2.08	3.43
Transport	0.00	0.14	0.38	0.62	0.97
Private Services	0.15	0.32	0.32	0.42	2.61
Public Services	0.29	0.32	0.40	-0.57	1.42
Macroeconomy					
CO_2	-3.11	-4.08	-7.99	-8.97	-8.24
Hard Coal	-7.97	-11.81	-14.85	-29.53	-24.87
Lignite	0.00	-0.02	-51.26	-2.10	-25.52
Oil	-0.30	-0.85	-2.56	-2.09	-0.75
Gas	-9.03	0.18	5.94	5.73	-1.52
GDP	0.13	0.26	0.35	0.45	2.34
Consumption	0.26	0.23	-1.20	-1.66	4.79
Investment	-0.13	0.87	1.14	2.07	3.90
Export	-0.19	-0.24	0.41	0.81	-1.71
Import	0.12	0.35	0.83	1.00	0.87
Employment	0.31	0.85	2.03	1.87	1.62
Capital Stock	0.19	0.49	1.15	1.59	1.87
Exchange Rate	0.03	0.00	0.00	-0.02	-0.10

Table E7b. Effects of European Carbon/Energy Tax under Alternative Tax Recycling in the EU (High Wage Elasticity)

	2000	2005	2010	2015	2020
Electricity Industry					
Total Generation	-0.02	-0.03	0.00	0.02	0.05
HC	-0.87	-12.74	-31.30	-40.55	-31.12
BC	-2.81	-0.17	-29.12	-26.49	-14.92
Oil	13.32	3.94	7.60	-8.98	-33.79
GAS	-2.98	22.75	57.77	61.38	42.47
NUC	0.00	0.00	0.31	0.00	0.00
Others	1.12	0.86	2.70	1.40	1.13
Imports	-6.50	-8.07	3.13	12.58	18.10
Exports	-6.63	-8.27	3.20	12.81	18.44
CO_2	0.22	-4.70	-16.45	-19.30	-12.78
Sectoral Output					
Agriculture	0.15	0.22	0.48	0.44	0.60
Hard Coal	-0.56	-7.31	-15.27	-19.29	-20.59
Lignite	-1.72	-0.80	-26.98	-15.78	-13.34
Oil	-0.21	-0.68	-1.33	-1.39	-2.13
Gas	-0.48	0.67	2.96	3.90	2.52
Electricity	-0.02	-0.03	0.00	0.02	0.05
Basic Materials	0.16	0.17	0.29	0.26	0.35
Equipment	0.28	0.54	1.08	1.00	0.94
Consumption Goods	0.30	0.43	0.92	0.80	0.98
Construction	-0.15	-0.02	-0.10	0.11	0.18
Transport	0.14	0.21	0.45	0.36	0.47
Private Services	0.21	0.27	0.56	0.42	0.58
Public Services	0.17	0.27	0.53	0.23	0.32
Macroeconomy					
CO_2	-1.07	-3.71	-8.17	-8.75	-7.11
Hard Coal	-2.00	-10.44	-24.30	-29.74	-23.00
Lignite	-2.08	-0.44	-22.93	-18.96	-9.56
Oil	0.28	-1.04	-2.58	-3.06	-3.79
Gas	-2.48	-3.24	-1.98	-0.69	-1.71
GDP	0.19	0.23	0.43	0.27	0.48
Consumption	0.20	0.12	0.27	0.03	0.25
Investment	-0.68	-0.69	-1.54	-1.12	-1.19
Export	-0.12	-0.16	-0.32	-0.25	-0.41
Import	-0.13	-0.20	-0.31	-0.29	-0.48
Employment	0.33	0.59	1.25	1.03	0.97
Capital Stock	0.05	0.07	0.17	0.13	0.11
Exchange Rate	0.03	0.00	0.00	-0.02	-0.10

Printing: Weihert-Druck GmbH, Darmstadt
Binding: Buchbinderei Schäffer, Grünstadt